VIRGINIA CLIMATE FEVER

VIRGINIA

HOW GLOBAL WARMING WILL TRANSFORM

CLIMATE

OUR CITIES, SHORELINES, AND FORESTS

FEVER

STEPHEN NASH

UNIVERSITY OF VIRGINIA PRESS | CHARLOTTESVILLE AND LONDON

University of Virginia Press
© 2014 by the Rector and Visitors of the University of Virginia
All rights reserved
Printed in the United States of America on acid-free paper

First published 2014
First paperback edition published 2017
ISBN 978-0-8139-3995-7 (paper)

9 8 7 6 5 4 3 2 1

The Library of Congress has cataloged the hardcover edition as follows:
Library of Congress Cataloging-in-Publication Data
Nash, Steve, 1947–
    Virginia climate fever : how global warming will transform
our cities, shorelines, and forests / Stephen Nash.
    pages   cm
    Includes bibliographical references and index.
ISBN 978-0-8139-3658-1 (cloth : alk. paper)
ISBN 978-0-8139-3659-8 (e-book)
1. Climatic changes — Environmental aspects — Virginia.
2. Global warming — Virginia. 3. Virginia — Environmental
conditions. I. Title.
    QC903.2.U6N36 2014
    363.738'7409755 — dc23   2014010471

For Celeste, Alex, Forrest, Jason, and Sanjar

CONTENTS

ILLUSTRATIONS

ACKNOWLEDGMENTS

Journalists set out to write about a subject like climate change as mendicants, relying on a deep fund of generosity from the expert sources they will ring up or visit with endless questions. I invite my readers to turn to the list of sources at the end of this book and know that I thank them all for their willingness to let me barge into their schedules, on multiple occasions and for long conversations.

Within that group, however, several rendered extraordinary service: the climate statistician Robert Livezey, Adam Terando of North Carolina State University, Chris Zganjar of The Nature Conservancy, Philip J. (Jerry) Stenger of the University of Virginia's Office of Climatology, James Titus of the EPA, Katharine Hayhoe of Texas Tech University, and Justin Madron and Tihomir Kostadinov at my own home port, the University of Richmond, to which I owe more than three decades of employment in teaching and in science journalism. Scott Zillmer of XNR Productions brought patient insight to the graphics; the fine wordsmith Chris Reiter added pace and focus. Some parts of this book first appeared in articles I have written for the *New Republic, BioScience,* the *Washington Post, Blue Ridge Country,* and *Chesapeake Quarterly.* I appreciate their support for my work.

I am happy to have this chance to express thanks for the long-term encouragement of Boyd Zenner of the University of Virginia Press, for the meticulous copyediting of Susan Murray, and for my candid circles of thinkers in the back room at Melito's in Richmond, especially my great friend Jim Bacon. That chorus of sturdy skepticism keeps me thinking hard.

I am also indebted, as are we all, to Thomas Jefferson, whose name turns up several times in these pages in one form or another. In an era of fierce hostility to science, he was its champion. That inspiration for our own good governance is profound, especially now.

Last and most, I owe the completion of this book to my wife, Linda Nelson Nash — tough editor, dearest comrade-in-arms, and most patient listener.

A CLIMATE CONVERSATION

It's just before sunrise, the moment when the Earth stops cooling but has not yet warmed, so the air is still. I can hear an approaching freight train as it mutters along in a slow crescendo, a few miles away. About a dozen trains come through here each day, and their burden is mostly fossil fuels: "bomb trains"—tank cars full of readily explosive North Dakota oil—or coal pulled east off the longwall seams and out of the rubble of the "mountaintop removal" strip mines of the Virginias.

Down at the crossing near my house, the rails arrow toward Richmond and the coast, and back the other way to Charlottesville and then the Appalachian coalfields. They're like a transect line—the string, marked at intervals, that biologists pull taut across a patch of land to sample it and take its measure.

I've been visiting many such waypoints across Virginia, places where climate research has gauged the likely impacts of future global warming. They are this book's destinations.

The tracks pass through the distant Blue Ridge town of Glasgow, for example, where they bisect the view from the porch of Julian Kesterson, a careful recorder of climate data for the National Weather Service. Other coal-haul rails cross the Alleghenies at the hamlet of Saltville, where the fossil remains of mastodons, giant beavers, and humans from the last ice age give evidence of the force of climate shifts on Virginia's living organisms, including us.

A few minutes pass, an air horn sounds a baleful blast at Gaskins Curve, and soon after that a couple of diesels locked to a hundred or so coal hoppers thunder past, rocking a little on the rails. Each one weighs about 143 tons loaded, and each chunk in those carloads is a miracle of embedded energy, created by photosynthesis and then some 300 million years of subterranean pressure.

Ignited, a pound of coal creates enough energy to power the computer I am using just now for about twenty hours. Americans use 33 pounds of it per capita on the average day, and about one-third the electrical energy consumed in Virginia comes from it.

On east of here, some of the run of coal I'm watching will roll into the waiting maw of an electric power generating plant. The one nearest where I live consumes 40 tons of coal an hour. As coal, oil, gasoline, natural gas, or other fossil fuels are combusted, their carbon atoms are released in carbon dioxide gas. That particular plant generates about 950,000 tons of atmospheric CO_2 a year.

The gas will waft gradually into the atmosphere and take up residence, some of it for centuries and about a quarter of it essentially forever — a hundred thousand years or so. That thickening global shroud of CO_2 and other greenhouse gases traps long-wave heat radiation reflected from the Earth's surface. Instead of passing through the atmosphere and out into space, more of the heat stays here, and that adds destabilizing energy to our global climate system.

Most of this train will make its way on through the Tidewater region along a route not far from the inland town of Smithfield. There, millions of years ago, climate change once pushed the Atlantic. The sandy ancient shoreline yields a climate story that is told in the chemistry of its fossil shells. Geologists and climatologists use it to help calibrate models that foretell global warming in our own era.

The tracks end at heaps of stored coal, dozens of feet high, grazed by giant cranes at a shipping terminal in Newport News. From there a conveyor pulls the coal like a black river, up into "bulker" cargo ships that will head out along the Atlantic seaboard or to Europe or Asia. Part of that terminal, where the mouth of the James River meets the Chesapeake Bay, may be inundated during the next few decades as global warming continues to accelerate sea level rise.

"Climate change," "global warming." I use the terms interchangeably because they're familiar, but "climate disruption" is far more accurate. U.S. National Science Advisor John Holdren has explained that "Global warming is a misnomer, because it implies something that is gradual, something that is uniform, something that is quite possibly benign. What we are experiencing with climate change is none of those things. It is certainly not uniform. It is rapid compared to the pace at which social systems and environmental systems can adjust. It is certainly not benign. We should be calling it 'global climatic disruption.'"

Our sources of energy and emissions change. The use of "fracked" natural gas, with its legacy of environmental damage and greenhouse methane, is also growing fast. New federal regulations for emissions will help — if they stick — although Washington also pushes aggressive fossil fuel growth that marches boldly backward.

Worldwide, the combustion of all fossil fuels — coal, oil, gas — has moderated, and growth per-capita energy use and global CO_2 emissions were flat in the most recent year they were tallied. They will determine the density of the greenhouse gas blanket we all live under.

CO_2 from fossil fuels is only one of several greenhouse gases, but it's the most significant, and it is plain that fossil fuels pervade our way of life. Just as clearly, the impacts of burning them are accumulating swiftly. Atmospheric CO_2 concentrations have risen from 275 parts per million (ppm) a couple of centuries ago, past the 350 ppm level judged relatively safe by some (not all) climate scientists, and above 400 ppm now. We are adding about 2 ppm each year. A recent headline summarizes the latest, most thorough international scientific study: *Climate Change Report Sees Violent, Sicker, Poorer Future.*

Every author anticipates his readers, though, and the next question from you could be: So what? Who's kidding whom? Because we have some big problems with talking about climate change. Number one is that — while the poll numbers shift easily, seeming to respond to each new round of snowfalls, hard freezes, and record-busting heat waves — maybe half of us don't believe that global warming is under way, or don't believe that humans have much to do with it, or don't think that it's much of a problem. And if we believe it this month, the odds are that a chilly year or two would dissuade many of us. "Global warming is still seen as a relatively distant threat," a recent poll found, "a threat distant in space and time — a risk that will affect faraway places, other species, or future generations more than people here and now."

If you're a citizen who tends to think the whole climate discussion is over-blown — some of my smarter friends do, too — salutations! I suggest you skip out to chapters 13 and 14 now for a conversation about climate models and credibility, and then decide whether it may be worthwhile to read further. If you turn out to be right in the future, you can have a more well-informed last laugh when the whole climate tempest passes. (If things don't develop as you expected, though, you may be in a better position to help the rest of us.)

Here's what I hope is a genial way to disarm the ferocious arguments we're all too familiar with, at least temporarily: maintain your convictions, but also listen to the scientists I've talked with. Judge their reasoning, evidence, and credentials, and compare them to the people you usually pay attention to. We can tune in any time to political pundits we like, and we can arrange to hear and read whatever version of reality pleases, on the Internet and elsewhere.

But what we really know about climate change — and how it will disrupt and alter Virginia's cities, shorelines, forests, and agriculture — doesn't originate with politics. It's science. Accordingly, one aim of this book is to let you listen in on scientists as they talk about their climate-related research, in nonscientific language. Where my sources of information aren't made explicit as you read, you can find them in the chapter notes.

So I won't pause to argue the point if your best thinking is that climate change is hysteria, a mistake, or a conspiracy. Nearly all climate scientists and others in related fields agree that it is well under way — as do most of the small caucus of "contrarian" scientists — and that human activity is its generating force. For science, that's a long-settled issue.

On the other hand, you may be a reader who's already convinced that climate change is real, and failing to address it is humankind's biggest and most dangerous gamble (nuclear weapons are an arguable contender). You might also know well that the United States, and notably, Virginia, have done little to help alleviate greenhouse gas pollution, or to prepare for its consequences. Virginia ranks seventeenth of the fifty states in per-capita energy-related CO_2 emissions. Americans more generally are easily among the planet's biggest greenhouse-gas polluters. Americans emit twice as much CO_2 as the average European and use twice the electricity of the average European or Japanese.

IF YOU ARE WORRIED about climate disruption, you may have concluded that we're in this nightmare jam because special interests have hijacked both our political system and our public conversations, in Virginia and nationally, by obfuscating and distorting what we have known about global warming for decades. After all, 40 percent of the increase in atmospheric CO_2 over its pre-industrial-era level has been put there just in the time that has passed since the first congressional hearings that warned about climate change, in 1988.

Alternatively, you may believe that nothing in human history or psychology has prepared us to be able to transform a carbon-based economy in so short a period, no matter the politics. If our civilization were that unstable and reactive, you might surmise, even larger problems would ensue — so we might as well push forward from here, as best we can. Either way, we're long overdue to start, in earnest, a conversation about preparing for the climate impacts that are irrevocably on the way.

Here's a potential problem with looking almost exclusively at Virginia. Even if you are convinced of the certainty of climate change, you've been taught for good reason to think of it as a global phenomenon. That concentration of greenhouse gases is more or less even as it wraps around the planet. Whether it originates in Virginia or in China, its impacts will be everywhere. Why does it make sense, then, to discuss the climate of one small patch of the Earth's surface, demarcated by our nervous-looking, oddly shaped state boundary, as if it were somehow separable from the climate of the rest of the globe?

For Virginians, it is useful because we live here and we make decisions together. We'll have to plan for climate change much of the time within the political confines of the state or of a community, rather than as part of some larger group. We usually identify ourselves as Virginians and not, except in the abstract, as citizens of the globe or of the Western Hemisphere. We have responsibilities to each other, to the natural systems we depend on, and to Virginia's landscape, one of surpassing richness and beauty.

We are also Americans, of course, deeply and directly affected by what happens in the rest of the country. As I write this, the United States has endured the hottest July in its recorded climate history. A brutal midcontinent multiyear drought has reduced the corn crop by 15 percent. Even so, America's climate is varied and mostly distant, not an immediately shared life experience. Virginia's climate, on the other hand, is what we wake up to every morning.

Another reason why it makes sense to talk about climate change on the scale of a single state has occurred only recently. When scientists first began to try to figure out what might happen in the future as a result of adding greenhouse gases to the atmosphere, their estimates were necessarily crude. They still usually work on a model of the Earth divided into a grid of trapezoids, each about 400 miles on a side — a picture so coarse-grained that even the Appalachians are only a smear. It is as if you were holding a five-by-seven photo comprised of pixels of about an inch square.

Those models, quite effective on larger scales, often cannot "see" much of the topography that drives local climate — the complex coastlines, mountain ranges, and valleys that affect heat and precipitation. Now, however, the maps of future climate can discern your region of the state. The grid unit of some projections is as fine as seven miles square.

Those maps are tentative. Their details sometimes meander and fade. The image is still, as it has to be, an assemblage of blurred fragments. But even with all the uncertainties, they are generally useful guides to the future, and on a scale we can more easily comprehend. You will read later about the disciplined reality checks that climate science uses to assure itself — and us — that the projections are not merely a tangle of shared delusion. With them, we can finally see the broad outline of climate change and some of its impacts in Virginia over the coming decades.

Whether we really want to look at that future, though, is not always clear. One of the clichés we've all grown up with is that knowledge is power. It's also one of the foundations of that most optimistic of enterprises, democracy. Let truth and falsehood grapple, John Milton wrote in his Enlightenment charter *Areopagitica*: "Who ever knew Truth put to the worse in a free and open encounter?" James Madison warned that trying to govern ourselves without good information "is but a prologue to a farce, or a tragedy, or perhaps both."

And yet we are also awash in cultural messages of the opposite kind, some everyday and some exalted. A desire to know the truth, we've always been warned, has its considerable disincentives: no news is good news, and ignorance is bliss. Eve in the Garden, Lot's wife, the Tower of Babel, Oedipus, Pandora, and the curious cats — they all testify on behalf of the beguiling idea that what you don't know can't hurt you — or at least that knowing hurts worse.

It is profoundly disorienting to think about climate change, which has been called a "slow-motion global catastrophe." Despite its adversities and wide variation, the present climate is the one you and I and our long lines of ancestry have lived in and adapted to. In the span of civilization, this is the climate our cultures have been shaped by. The changes forecast by climate models could soon take the planet back to a much hotter time millions of years ago — deep time, so long before our species evolved that it is nearly beyond fathom.

The options we have, as Holdren has pointed out, are to take steps to reduce the amount of climate change we're causing, to adapt as intelligently as possible to the change we can't avoid, and to suffer. "The question — the issue that's up for grabs — is what the mix going forward is going to be," he has said.

It seems plain that in this case we'll be far better off knowing than hiding. Climate change is already upon us in Virginia and everywhere, but we still — and this is our best hope — have time to work out good plans in the face of it, and avoid making the worst of it.

VIRGINIA'S CLIMATE NOW

Some of Thomas Jefferson's *Notes on the State of Virginia,* published in 1782, sounds more than a little familiar: "A change in our climate however is taking place very sensibly," Jefferson wrote. "Both heats and colds are become much more moderate within the memory even of the middle-aged. Snows are less frequent and less deep. They do not often lie, below the mountains, more than one, two, or three days, and very rarely a week. They are remembered to have been formerly frequent, deep, and of long continuance. The elderly inform me the earth used to be covered with snow about three months in every year."

Jefferson was a faithful and meticulous recorder of weather, but as critics of the time noted, his annals were still too brief and too local to substantiate a change in the climate, and he put too much stock in the anecdotes and suppositions of those weather-watching elders. Virginia's records are longer and more comprehensive now, thanks to a National Weather Service network of cooperative observers — exactly the kind of system Jefferson advocated.

Julian Kesterson, a lifelong resident of the little town of Glasgow in Rockbridge County, has been one of those volunteer observers for more than half a century. He greets visitors warmly but also with an appraising gaze from behind steel-rimmed spectacles. He speaks of climate not from memory but with the authority of his instruments and logbooks.

"I started keeping records on the first day of January 1960 for myself, and then I started being an official observer for the Weather Service on the first day of September 1967," he told me in a measured mountain-Virginia twang.

His porch has a fine, broad view out over a double set of railroad tracks and then the James River, a couple of hundred feet away. The irony of the passing coal and crude oil trains is not lost on him.

"In the years I've been doing it, it's definitely gotten warmer," he said, reeling off a string of temperature data from his files. "I mean, they can argue all they want to, but I have the records to prove it's really warmed up here. There's no question about it. It's kind of scary. And you know what really concerns me is so many people make fun of these scientists, and claim they don't know what they're talking about. I mean, I just can't understand it.

"Temperature has gone up, and the rainfall has gone up slightly over the last fifty-three years that I've been keeping records. About three inches more a year on average than what it was, years ago. But the temperature's been the big thing, and the winters have been the most different." That trend jumped abruptly one recent winter — the warmest he'd recorded. "I really wouldn't call this a winter. It's more like a Twilight Zone," he told a reporter that February.

That's how the climate trend looks from Julian Kesterson's place, a sturdy white frame house with a green metal roof and an arsenal of instrumentation in the yard. It is hemmed in by the 700-foot walls of the James River Gorge near its confluence with the Maury River, 737 feet above sea level and about a mile southeast of the town. His data has great value, but its particularity raises obvious questions: how representative is it? Is it evidence of a trend toward more warming, or just natural variation?

THE ANSWERS CAN BE found in a broader and deeper record of the climate of the Southeast. It has been relatively stable since about eight thousand years ago, but that doesn't mean that it has been uniform. Virginia's place at the edge of the continent and the varied terrain across its 462-by-201-mile expanse have given it a remarkably diverse climate, compared to states that are uniformly flat or dominated by either mountains or oceans.

Annual precipitation totals 35–37 inches on average in much of the Shenandoah Valley, for example, but rises to more than 55 inches in the mountains of Southwest Virginia. Winter temperatures in the higher elevations of the northern Blue Ridge have sometimes plunged into the same deep freezes as Chicago on the same days when Western Piedmont farms, not far away, see much milder temperatures.

These contrasts are mostly the result of contention among rivers of air that join or collide over the state, each bearing its own temperatures and humidity. A suite of factors determines their character.

Along the Atlantic coast, the warm, northward-moving waters of the Gulf Stream moderate winter temperatures. During storms, onshore air currents can push warmer, more humid coastal air inland toward the eastern face of the Blue Ridge, which at times wrings drenchers and occasionally catastrophic floods out of the trapped, moisture-laden clouds.

When air currents flow eastward over the state, the Shenandoah and New River valleys stay relatively dry, because they are in the "rain shadow" of the Alleghenies that flank their west side. When weather moves the other way, to the west, the Blue Ridge can bar much of the rain from reaching these valleys, too.

The James, Shenandoah, New, Roanoke and other river valleys also function as conduits that direct lower-altitude air currents. If the currents are moving from lower to higher elevations along those valleys, rains can increase during their journey. If they're flowing down-valley instead, the chance of rain tends to taper off as they descend.

"Climate is what you expect, weather is what you get," the saying goes. Another way to look at the difference between them is that weather is what's happening outside today, but climate is the average of what has happened there over time — say, thirty years or more. Climate is the long average of weather.

The National Oceanic and Atmospheric Administration (NOAA) designates six different climate zones in Virginia, each of them responding somewhat differently to the fixed bumps and hollows of the state's own topography and the swirling flow of continent-scale climate factors. The boundaries of these climatic zones look sharply defined on paper, but they're just abstractions — summaries of the flux of weather patterns through time. As the Virginia Office of Climatology cautions, "A climate condition typical of one region might in a given year extend outward into another area."

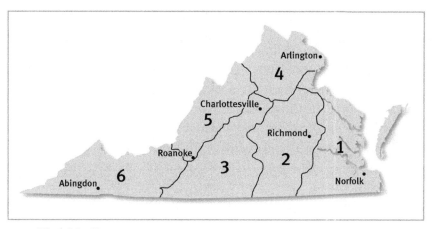

FIG. 1. Virginia's climate zones

The map of Virginia's climate zones reflects historical data, meaning that it looks backward to the patterns of the past for insight into what may happen in the future. That works well enough in a generally stable climate. It has always been a valuable enterprise for agriculture, and for the rest of us.

Increasingly, though, historical data do not have the same value for predicting seasonal and geographic climate patterns. Those data trends are less cyclical now, more in transition. We are moving out of the patterns to which we're accustomed and into variations that will be new, at times radical, and not as predictable. We don't know where they will ultimately lead until a new climate equilibrium arrives, perhaps centuries from now or longer. If the climate is changing, our expectations have to change, too. In that case — our case! — it's a set of rolling averages, defined by how far back in time we're looking.

Some of that evolving story can be heard at the offices of Jerry Stenger, director of the Office of Climatology at the University of Virginia, where he has worked for more than thirty years. On a recent afternoon he directed some graduate students, fielded phone calls from reporters around the state about the exceptionally warm winter (soon to be followed by an exceptionally cold one), and mulled over a lecture he'd just heard by a visiting climate scientist.

I asked for the names of the climate change specialists on his staff. Stenger tilted back in his chair and smiled, with teeth. "Well, let me put it this way," he said. "I am devoting the entire effort of everybody in the Climatology Office to you, right now." He is the whole deal. (Different states handle this, predictably, in different ways. The North Carolina state climate office at North Carolina State University, for example, has a dozen full-time staffers and nine additional graduate and undergraduate assistants.)

The real action in weather comes with winter, Stenger explained. The equator and the tropics get a comparative surplus of energy from the sun during winter, but the poles have a deficit, and the imbalance tries to right itself. The key to weather, then, is moving the equatorial energy toward the poles. "That's what weather is all about: transferring that energy," he said.

One complicator during the transfer is that the Earth spins. As it does, a fast-moving boundary develops between the warmer equatorial air and the colder polar air. That boundary is the jet stream.

A half century ago, a couple of meteorologists re-created the jet stream effect in a clever and quickly famous experiment that used a round, flat-bottomed dishpan full of water. The dishpan was intended as a sort of two-dimensional climate model of the planet, as if it had been squashed flat. The water stood in for the atmosphere. The North Pole of this atmosphere was at the dishpan's center, which was cooled by the experimenters as part of the simulation. The equatorial zone was represented at the outer edge, so they heated that.

The whole dishpan was put on a turntable made to spin once every three seconds — mimicking the Earth's rotation. In between the colder water and the warmer, a boundary materialized as the dishpan revolved. It was a wavering stream of fast flow — a jet stream. That fast-track dividing line moved farther from, or nearer, the "equator"—the outer edge of the dishpan — depending on the difference in temperatures between the heated equatorial zone and the cooled polar center.

That's what happens in the Earth's atmosphere. The fast-moving boundary, the jet stream, shifts north and south, depending on how much difference there is between equatorial and polar temperatures. North of the jet stream, it's colder. South of it, warmer temperatures prevail. The jet stream is S-shaped over North America, especially in winter, with a north-bending ridge in the west, and a south-bending trough in the east. The summer jet stream is less wavy, much slower, and moves farther north as a whole.

Virginia's horizontal shape often lies across that winter boundary and its undulations and eddies. "We're kind of in the cross-hairs, north-south and east-west," Stenger said. "It makes winter forecasting very difficult in this part of the country." A 12- to 20-mile distance can make the difference between a big snow in Charlottesville or Fairfax County, and rain in nearby Zion Crossroads or Woodbridge.

The continent's topography and coastal temperatures tend to lock the general jet stream pattern more or less in place, though on occasion there is a reversal. "They can get snow in Arizona, and in the flatlands of West Texas, and a lot more rain in California. In Virginia, we'll be walking around in our shorts in January," Stenger said. "But it doesn't stay that way for long. That's why they drive more convertibles in California than here."

Three days before a recent Christmas, the tree in my backyard that usually blooms in early April bore exuberant pink buds under the provocation of a string of days in the 70s. As Stenger is careful to point out, however, a single heat wave, no matter how remarkable, no more represents a conclusive argument in favor of global warming than the bitter, record cold that followed ten days later — it froze buds, birdbath, and ditches — was evidence against it. Climate isn't an episode, it's a long series. "We can't draw conclusions from these events, one way or the other," he said.

So if you don't accept that climate change is under way, no particular day's or season's weather can be a convincer. When you hear climate scientists occasionally hesitate and even disagree about whether to attribute one or another drought, flood, or violent weather event to global warming, it is because no single short-term phenomenon can be labeled that way, and the longer trend may not be clear.

They can, however, draw conclusions from sustained patterns over larger areas and longer periods. The more emphatic those patterns, the stronger the proof. The winter unfolding outside Stenger's office that day was the fourth-warmest on record for the contiguous United States. But much more significantly, these temperatures were a part of a larger, lengthening pattern. The 328th consecutive month with a global temperature above the twentieth-century average soon followed. The last below-average-temperature June for the United States was June 1976, and the last below-average-temperature month was February 1985. As of that July, 56 percent of the contiguous United States was suffering through a severe, and lengthening, drought.

So climate scientists point out that although climate disruption isn't dictating one or another day's or season's weather, it is instead increasing the odds that set the pattern and direction over time. A 73-degree January day was always possible in Richmond or Roanoke or Fairfax a long time before global warming hit the news. But it is more likely now, and will become likelier still in the future.

Stenger led me up to the McCormick Observatory, a brick octagon of pleasing proportions on a secluded green hill nearby. It was built for astronomy in 1877, but is also the home of one of the state's longest-running weather observation stations.

He pulled a battered old ledger out of a desk drawer, an official weather record whose entries date back to 1871. (Contemporary records are also kept with a keyboard connected to cyberspace, but pen and paper are still in use — the final arbiters if questions arise.)

"If we're going to look for human influences, we should really start in about the mid-1970s," Stenger said. "That's the time when the world economy got off to a big start, and things began to grow quickly. We began to have satellite data, so now we have about thirty-five years or more of trend line to look at — records we consider as very good indicators of what's going on."

Global climate change will present itself as large-scale changes in atmospheric circulation, and there are already some areas around the globe where temperatures are increasing rapidly — what you would expect to see, based on the release of greenhouse gases. "There's no evidence of any process that's likely to change the trajectory we're on, as far as the warming goes," Stenger added.

Most areas will undoubtedly warm when the circulation patterns change, but at least some may get cooler. Counterintuitively, some areas could get larger snowstorms as the patterns of atmospheric energy change. In most others, snow and ice will become rarer, or will disappear. There is general agreement, Stenger said, that we are seeing a significant trend: global temperatures have gone up overall, along with those in the United States.

But those are averages over large areas. The smaller the geography you look at, the more variation there is and the more uncertainty you find. It is not clear-cut without further analysis (which will appear in a few pages), how much of Virginia's warming can be attributed to a global climate shift, and how much may be within what we expect from natural variation.

Derek Arndt, chief of the Climate Monitoring Branch of the National Climatic Data Center, told me that this is like linking how children are raised to their behavior as adults. "It's a long-term influence that affects almost every outcome pervasively, but over time, you really can't connect good or bad parenting to any single decision that a person makes. If I go out and rob a string of banks, the things that would drive that decision would be that I need money, it was easy to do, I formulated a plan. The parenting I received would have surely been a part of that decision process, but pinning any horrible individual single decision on parenting is just really difficult."

Similarly, as you consider smaller, Virginia-sized parts of the generally warming continent, a large number of local and immediate factors play a lot bigger role in outcomes. "As we carve up the world into smaller and smaller pieces, they begin to show their own histories and the global temperature trend is a composite of all of those histories," Arndt said. "The warming isn't monolithic over time or space."

It's the overall "parenting" role of background climate change that registers on Virginia over time, not the mistaken notion that climate change is the main force that twists the control knobs from day to day and from season to season. "Climate change is the pervasive influence that expresses itself in ways that we are still catching up to understanding," Arndt said.

Stenger agrees that only a fraction of what we've seen so far in Virginia is necessarily attributable to climate change. As he puts it, though: "That things are changing and will continue to change is definitive — that's certainly the way to bet. It's going to get warmer in Virginia." The trend lines can still jerk up and down in the short term. You might lose your money on an annual bet on this trend, he said, but you'd win if you bet on a decade.

Seeing the trends more clearly means dimming the influence of those shorter-term variations in the data. Arndt explains the first and easiest way to clarify: "Sixty-month averages are going to get rid of a lot of year-to-year

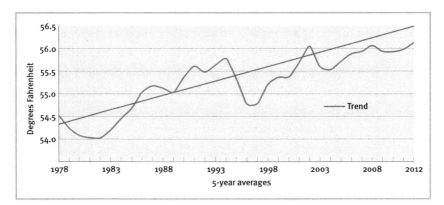

FIG. 2. Virginia temperature trend since 1978

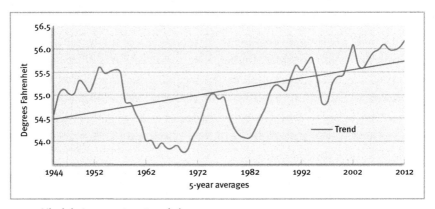

FIG. 3. Virginia temperature trend since 1944

noise. You are going to be left with something that more represents the long-term signal. It would be as if I graphed my weight over time. I have had some years when it's been up, some years when it's been down, but as I get older, the long-term signal is up." A five-year average isn't deflected by, "Oh, that was that summer I decided to work out."

When I began to ask about the trends in Virginia's climate record, I learned that plenty of numbers are out there but interpreting them can be a puzzle at first. How much warming we see during the past decades depends a lot on what data we choose to look at.

Even if you are mildly allergic to charts, the ones in this chapter should decipher easily: The wavy blue line in figure 2 shows the variation in sixty-

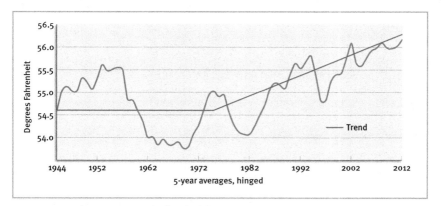

FIG. 4. Increasing temperatures in Virginia

month average temperatures since 1978 for the whole state, and the straight line running through it summarizes the upward slope — the general trend — over those thirty-odd years.

Though it may not look startling at first glance, that is a very steep rate indeed: about 6 degrees of warming per century. (I will use Fahrenheit throughout this book.) There's a problem, though. Starting in the 1970s doesn't take into account the fact that they included some of the coldest temperatures in the whole century. The result is a kind of optically exaggerated warming trend. If added CO_2 is heating things up, that heat should have been evident quite a while ago. But if we extend the trend lines back to 1944, as in figure 3, the rate of warming is far less — only about 1.8 degrees per century.

That statistical tailoring doesn't explain what has really happened either, though. It downplays the recent warming trend disproportionately. In order to see the real picture, we have to take both the steep 1970s trend and the data from the 1940s forward into account, but within their proper context.

THE NATIONAL CLIMATE DATA CENTER (NCDC) referred me to Robert Livezey for help in getting a truer picture of recent temperature trends in the state. He is an expert consultant in the world of climate statistics and was chief of Climate Services at the National Weather Service during a thirty-seven-year career, specializing in climate variability, change, and prediction.

His quality control process is worth a look. The first step was to collect all

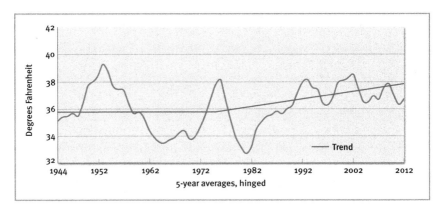

FIG. 5. Winters in Virginia: Dec.–Feb. trend since 1975 is 5.5°F per century

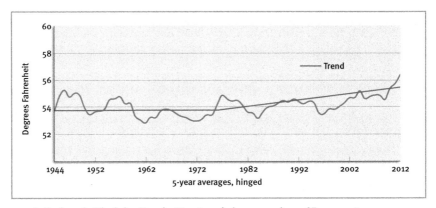

FIG. 6. Springs in Virginia: March–May trend since 1975 is 4.7°F per century

the Virginia temperature data from NOAA in each of the state's six climate zones. Then he quickly trashed the subpar, B-list weather stations whose data are suspect. Maybe urbanization and its hard surfaces swamped the station site over time, skewing the numbers in a warmer direction, or the instruments were relocated at some point, or there were inconsistencies or gaps in the reporting. Instead, he relied on a premium list of stations called the Historical Climate Network. Their data have been subjected to what he called "a number of carefully thought out corrections and adjustments" by NOAA.

Next, Livezey applied a statistical technique called a hinge-fit analysis to the data. It is a cumulative average, which avoids the false trail that is laid

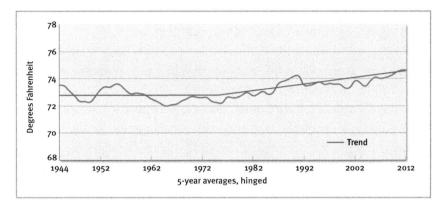

FIG. 7. Summers in Virginia: June–Aug. trend since 1975 is 4.9°F per century

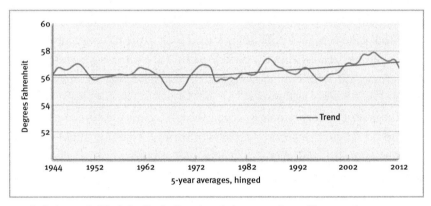

FIG. 8. Autumns in Virginia: Sept.–Nov. trend since 1975 is 2.7°F per century

when you ignore everything before the 1970s, and also accounts for the steep warming trend since then, but not disproportionately. Finally, he double-checked his analysis with NCDC.

What you're looking at in the trend line in figure 4, then, is the change in Virginia's annual average temperature between 1975 and today that shows the influence of climate change.

Natural variability is not part of the new heat, Livezey said. "This focuses us on the warming that is associated with large-scale climate change." The slope of his trend line — confirmed by the state Climatology Office through 2015 — shows that the statewide rate of temperature increase has averaged .46 degrees per decade since 1975, or 4.6 degrees per century. Scientists agree that as it continues, it ratchets up the risk of dangerous, destabilizing, and very long-lasting climate change within our lifetimes.

The steepest warming trend during that period has been in Division 3 — the southern Piedmont, in the lee of the Blue Ridge: .53 degrees per decade. The slowest is in Division 5, the Shenandoah Valley: .39 degrees per decade. For each of Virginia's climate zones, the trend of winter temperatures since 1975 has increased markedly, just as climate models predict, while summers warmed mildly in some zones and sharply in others. The statewide averages were .55 degrees of warming per decade in winter, .49 in summer. Autumn had slower warming, .27 degrees per decade, and spring, .47. You will find those seasonal trends charted in figures 5, 6, 7, and 8. (For local data on rising temperatures by region and city, go to virginiaclimatefever.com/5-2/localdata/.)

So that's the clearest available backward look — the recent record of Virginia's climate trend within the general picture of a warming planet. It's getting hotter. "What the data show is consistent with long-term climate change in Virginia," Livezey told me.

Of course, it's not as simple as just pushing Virginia's historical trend line on into the future to forecast the rate of warming. That line might conceivably bend upward, making the heat gain more rapidly, or we can even imagine that it might flatten or trend down. We have to rely on climate modeling to help us estimate the likelihood of those developments, and we'll see that they portend clearly what the recent trend suggests: more heat.

Four or five degrees' difference in temperatures may sound negligible. That's a fraction of the commonplace variation between daytime and nighttime temperatures, wherever you live in Virginia. The real impacts of those few degrees of change in the average are anything but trivial, however. A look forward along that projected trend line, and its potential consequences for the well-being of the forests, the cities, the Chesapeake, the Atlantic, and ourselves, comes next.

BACK PORCH, FORWARD VIEW

I once attended a workshop for journalists at the National Center for Atmospheric Research in Boulder, Colorado. I listened while an earnest physicist explained the mainsprings, cogs, and calipers of the various climate models for North America. A reporter for a small paper in Bay City, Michigan, finally raised his hand and asked one of those simple but incisive questions: "What's it going to look like when I step out onto the back porch in forty years?"

Exactly. We want to know how hot it will get, how wet or dry, how much extreme weather there will be, and when all of it can be expected to arrive. Climate scientists can only tell us some of those things, however, and only in rather more general and tentative ways.

We can reframe the reporter's astute question: Instead of forty years from now, how about within the range of thirty to fifty years? Instead of how hot will it be, how about: How hot will an average summer be, over that longish span of time, as forecast by several different models? Instead of "how much will it rain?" we could ask whether we have been able to determine the general direction — more rain or less rain — that precipitation will take. And whether it will arrive in the form of bigger storms and heavier downpours.

There is a prior question, though: Why should we regard models as something other than a kind of mysterious received wisdom? How does the whole modeling enterprise escape the dread hazard of "confirmation bias," in which the research filters out any reality that doesn't conform to a preconceived notion (such as "global warming")? These issues involve research strategy — what models get right, and how we know it's right.

The climate models in use today, assembled and refined over decades by separate teams of scientists around the world, generate somewhat different forecasts. That may seem confusing, but it is a strength of the way the work is pursued. Where the models overlap, we can be more certain of their findings. Where they differ, we can bracket the uncertainty, and distinguish likely, unlikely, and nil possibilities. This is especially important for regional modeling, because the smaller the area — Virginia, say — the broader the uncertainties.

One common test is what happens when a model is programmed to "run" the climate we know best, that we have kept records of for the past 150 years. The projections of climate models in common use re-create closely the general pattern of historical temperatures over the land and the oceans, continent by continent and globally. The modeled paths of rising land and ocean temperatures, considered separately, are also close to observed reality for the century.

We can think of all this as an instrument, a box with a couple of tuning knobs, each knob representing a choice — a set of assumptions, really — that the operator has to make. First: the rate and degree of climate disruption will depend most on how quickly and completely we can rein in greenhouse gas emissions, if indeed we can at all. Climate modelers dial in a scenario, an assumed level of CO_2 emissions, and each setting generates a different range of results.

I am calling one such setting the "work and hope" scenario — a phrase you won't find in any federal report or science journal article I know of, but one that seems apt. It is presented in more detail in chapter 12. A worldwide crash program to halt CO_2 emissions may result in economic convulsion — or, new International Monetary Fund research says, it will save trillions and be fairly easy. (More than one-third of Germany's energy now comes from renewable sources — why not us?) Either way, it's what we need in order to slow the building heat and avoid more climate chaos ahead. It is the climate researchers' best-case scenario.

The other scenario you find in this book is one of those referred to as "business as usual." It assumes that emissions will continue to increase through most of the century, with the continued growth of world populations and the burning of fossil fuels and forests. In figure 9, the blue line

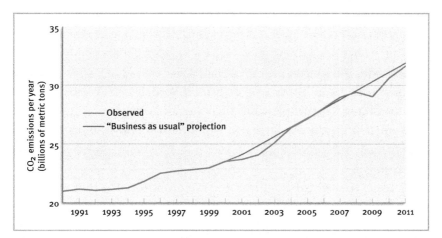

FIG. 9. Global CO_2 emissions are tracking the "business as usual" scenario that climate modelers have used.

shows actual global emissions during the recent past, and the other line is a "business as usual" emissions scenario used in modeling. The economic recession caused the recent dip in the emissions rate you see on this chart.

The second tuning knob on the modeling box selects which climate models you decide to use. You could consider them separately, for example, or pick a group of them that defines a range of future climate results. Or you could blend several and average their output.

The maps and other projections included in this book agree in general but not always in detail, because they are based upon different groups of climate models. Some models project that the climate is more reactive, or sensitive, to additional concentrations of greenhouse gases. Other models project that the atmosphere responds less quickly.

None of the maps of Virginia's future climate you'll see is likely to be entirely correct. Put another way, this is the best information that science can bring us for now, but it more resembles well-informed handicapping than truth telling. This "truth" is giving us a set of probabilities, not a unitary prophecy.

So the projections can't tell us about the weather off the back porch in May 2055, but they offer the range of possibilities. We can at least begin to think through, and plan for, the consequences within that range. No matter how the climate modeling box is tuned, global warming and its associated disruption occur. They just arrive years or decades earlier or later. Controlling greenhouse gas emissions means we warm more slowly, and the climate stabilizes sooner.

Those changes in future temperatures may look modest, benign even. Because averages flatten the daily highs and lows, they tend to camouflage what will be a rapidly changing new world whose outlines begin to emerge in subsequent chapters. For now, take a look at some projected average increases.

Figure 10 was created by Chris Zganjar, a scientist at The Nature Conservancy. It combines the projections of sixteen different models to approximate how far Virginia's summer climate could move down-latitude by around 2050 and 2100, given a "work and hope" (orange) and a "business as usual" scenario (red).

According to these projections, if we continue more or less on a "business as usual" path with emissions, then Virginia will be as hot as South Carolina sometime around 2050, and as hot as northern Florida by about the year 2100. Beyond then, the average annual temperatures of the new "Virginia" will continue to move beyond something like the latitudes of the Deep South, and then on into tropical zones.

People thrive in such places though, don't they? Coconut palms and cocktails — how daunting could that be? More short-sleeved shirts! Unfortunately, Virginians won't have the advantage of bicoastal sea breezes to counteract the heat. It will be more like laying Virginia into the swelter of interior Alabama and Mississippi, and then pushing that whole land mass farther south as time passes.

Even so, you could try to make this case stronger, the Duke University forest ecologist James Clark muses. "If you lived in Saskatchewan or something, you might say, 'Well, why wouldn't I like global warming?' There's no question that places that are pretty uncomfortable now could become more livable.

"But there's an awful lot of disruption to natural ecosystems that we can expect with that sort of rearrangement. The destabilization of the climate system is really, really scary." As we'll see, the pace of the transitions may turn out to be faster than natural systems such as forests can keep up with.

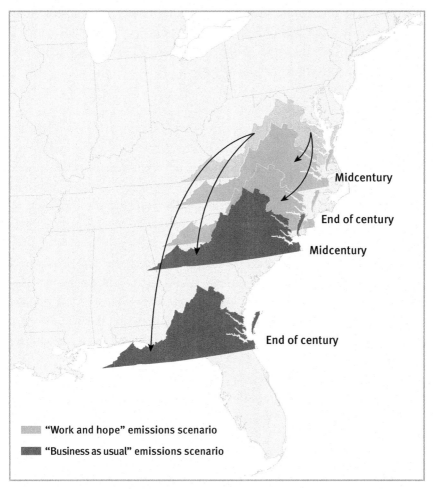

FIG. 10. Virginia's migrating climate

And although the chart lines show a steady trend, Clark said, in reality the climate may well lurch and wrench its way forward.

"You can't just sort of say, 'It'll get a little warmer everywhere and we won't mind that.' The warming will change the cities we live in and the forests, agriculture, and waters we depend on in ways we may be forced to adapt to, on an abrupt schedule.

"It's the climate that makes the globe livable," he said. "We can make a life here only because we've got this life-support system. Just more warming is not all that's going to happen. It's going to translate to new climates that we don't expect." Or as the climate scientist Katharine Hayhoe has summed up, "The outlook for the future is increasingly bleak." She is hopeful just the same that humans will respond, and cope.

HAYHOE IS DIRECTOR of the Climate Science Center at Texas Tech University and an expert reviewer for the Intergovernmental Panel on Climate Change (IPCC), which I refer to frequently in these pages. Founded in 1988 by the United Nations and the World Meteorological Organization, the IPCC has been the preeminent international research review organization, maintained through the voluntary participation of thousands of climate scientists.

Hayhoe has also contributed important research for a U.S. National Academy of Sciences report on climate stabilization, and for the most recent U.S. *National Climate Assessment,* a quadrennial report to Congress and the president that is mandated by law. She is among the originators of the statistical techniques used to "downscale" climate models — the basis for our regional picture of climate change impacts.

And just to jostle some wobbly stereotypes: she is an outspoken evangelical Christian, married to an evangelical minister. They are coauthors of the book *A Climate for Change — Global Warming Facts for Faith-Based Decisions.*

"Having a relationship with the God of the Universe is one of the most incredible experiences that a person can have," she has said. The hardest part of being a scientist and a Christian, she adds, is the amount of disinformation about climate science being targeted at her Christian community.

I have recited these credentials because at my request Hayhoe and the geospatial scientist Sharmistha Swain, also at Texas Tech, assembled figures 11, 12, and 13, depicting Virginia's future climate with the most recent and most refined available projections. From the larger number in use around the world, Hayhoe worked with eleven climate models, carefully chosen and extensively tested. Their projections were statistically downscaled to produce the regional picture, which was then evaluated with further quality-control tests.

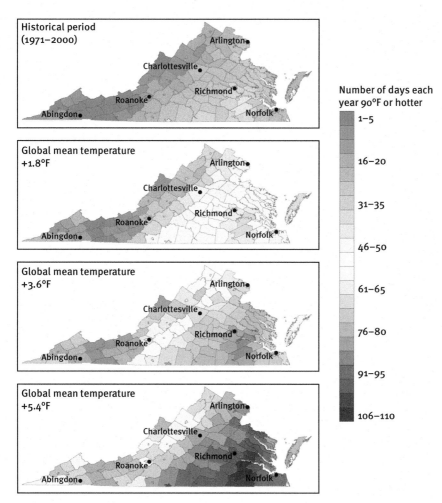

FIG. 11. How many more days that are 90 degrees or hotter will Virginia see as greenhouse gases intensify?

These maps offer our clearest picture so far of how the gathering global heat is likely to be distributed across Virginia. They are keyed to three incremental increases in global mean temperatures. As global mean temperatures are driven 1.8, 3.6, and 5.4 degrees F above those of the period 1971–2000, we can see what is likely to happen in different parts of the state. (In the metric system used by scientists, these increments are the cleaner-sounding 1, 2, and 3 degrees Celsius.)

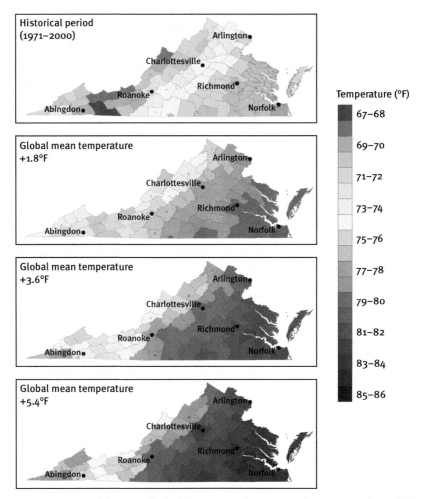

FIG. 12. How much hotter will Virginia summers be as greenhouse gases intensify? (Mean daily temperatures)

The maps show the effects of climate change on temperatures county by county, but Hayhoe cautions that they should be viewed less precisely, as projections for small clusters of counties rather than for just one. Also, the maps combine and average those eleven climate models, but of course the individual models' projections differ. So within the averages is what we might call a "range of uncertainty"— the plus-or-minus spread between the highest and lowest projections. That range increases the further into the future we look.

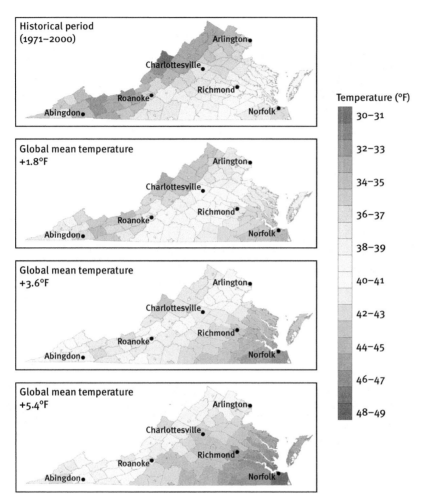

FIG. 13. How much warmer will Virginia winters be as greenhouse gases intensify? (Mean daily temperatures)

As an example, let's look statewide instead of at clusters of counties as the maps do. Given a 3.6-degree mean temperature increase for the planet, the highest projection among Hayhoe's models shows a total of 71 days above 90 degrees each year across Virginia, compared with only about 20 such days in recent times. The lowest projection is for 33 of those 90+ days across the state.

The range of uncertainty across the models, then, is 38 days. It's a substantial spread, but even the lowest projection — the model that shows a climate that is the least sensitive to greenhouse gases — tells us we are in for plenty of additional heat: the number of 90+ days each year across the state increases by 65 percent.

We don't know what the world will do about greenhouse gas emissions, or how fast world population will increase. We don't know except in somewhat broad terms how sensitive the climate will be to those emissions — that is, which models will prove most accurate. Climate scientists are certain, though, that the higher the rate of global emissions, the more rapidly Virginia temperatures will resemble these maps.

In the "business as usual" emissions scenario we have used, midrange models project that global temperatures will climb an additional 3.6 degrees circa 2050. That's the outer margin of what scientists hope is not catastrophic. It reaches 5.4 degrees by around 2065.

In the "work and hope" scenario, global mean temperature reaches +3.6 degrees, but only well past midcentury. Then it begins to level off as 2100 approaches.

You can find your city or county on the maps and assess the result. Where I live, in Richmond, adding 1.8 degrees to global mean temperatures means 61 to 65 days a year with temperatures over 90 degrees. That is three weeks more of 90+ days than I've experienced each summer, on average, during three decades of living here.

At the +3.6 degree mark for the planet, Richmond is likely to experience yet another three weeks above 90 degrees, for a total of 81 to 85 days, or twelve weeks. When we get to +5.4, Richmond could experience nearly a hundred days a year over 90 degrees, on average.

Figure 12 shows how climate disruption heats up twenty-four-hour temperature averages in future summers. With +1.8 degrees globally, Arlington's summer average advances by three degrees. At +5.4 globally, yet another 4 degrees will be added to summer averages in this part of the state.

Figure 13 projects winter climate averages that warm up Norfolk by 2 degrees, then another 2 at the next level, and another 1 at the +5.4 mark.

Missing from the three maps, because they are averages, are the spikes in heat that higher overall temperatures will bring. One impact of these is that when the heat comes, it is even hotter among the hard surfaces and breezeless canyons of our cities, especially around the asphalt. Heat is trapped, retained, and amplified by this "urban heat island" effect, to which the sick and the elderly are the most vulnerable.

Studies of past heat waves in the United States, Europe, and Asia have consistently found hundreds to thousands of heat-related premature deaths among these classes of victims, a phenomenon picturesquely referred to as "harvest." From 1979 to 2003, excessive heat exposure contributed to more than eight thousand premature deaths in the United States. That exceeds all deaths from hurricanes, lightning, tornadoes, floods, and earthquakes combined, during the same period.

More southerly cities adapt to these heat waves, of course, and so will we. Understatement of the era: a saner, cheaper, and more humane adaptation would be to forestall as much of the heat as possible, by working out how to cut back on the world's output of greenhouse gases.

HEAT IS EASIER to project than precipitation. Heat is regional, but rain and snow are far more local. Rainfall is more affected by topography, and thus more difficult for climate models to cope with. Cloud systems are the basis of precipitation, and for now, climate models can simulate clouds in only the most general way. The global maps of future precipitation produced by the IPCC have always labeled the Southeast region, including all of Virginia, a "zone of uncertainty."

How much rain or snow falls on Virginia, and when it arrives, are crucial for agriculture, water supplies, flood control, and a long list of other support services we've contrived for ourselves over the centuries. But in our region, the complexities of projecting precipitation in the face of climate change are, for now, beyond our ken.

Take a look, in figure 14, at two projections for rain and snow in Virginia in the year 2100, both using a high-emissions scenario and two different climate models.

One shows an increase in precipitation, compared with the years 1961–1990, of as much as 30 inches and at least 18 inches across the state. But the other model projects a decrease of as much as 20 inches, holding all factors in the comparison constant. These two models are at the extremes of "dry"

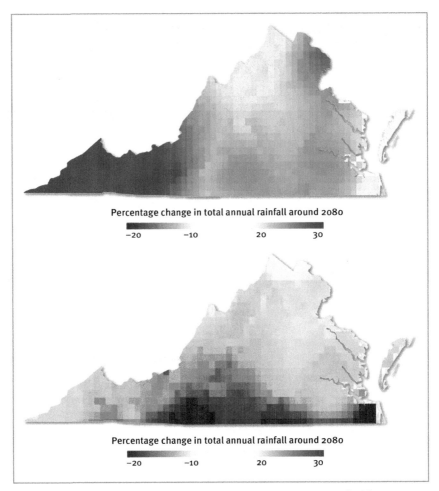

FIG. 14. Change in Virginia annual rainfall around the 2080s compared with current averages, using a "business as usual" emissions scenario and the "wettest" climate model (*top*) and the "driest" model (*bottom*)

and "wet," to illustrate the range of uncertainty. As you can see, they flatly contradict each other about rain in Virginia.

Duke University's James Clark explains that one of the tricky elements of predicting precipitation here is the Bermuda High, a pressure system that sets up offshore during the summer. "In the case of the Southeast, the clockwise circulation around such high-pressure systems spins moist air onto the eastern continent and is responsible for much of our summer precipitation," he said.

"What we've seen is that the precise location and strength of the Bermuda High can have a big impact. It's inherently unpredictable, relative to other parts of the globe, and it makes the Southeast a hard place to model. We've already seen quite a large increase in the variation in the precipitation here, but this is one of the places where models of the atmosphere just disagree on what the future climate will be."

In a couple of ways, the disagreement may be less important than it seems. Even if the rains increase, the onset of high temperatures over Virginia could cause most of that moisture to evaporate. And the rains are projected to arrive in comparatively short, intense storms — more extreme weather events, in other words — rather than in the steady train of more moderate storms we are used to. So they will not benefit us as much as they have in the past. This is especially significant for agriculture.

Assume that average rainfall figures for the years around 2020 turn out not to change much, all the way out to 2087. Conditions may be starkly different, just the same. Moderate rainfall in adequate amounts, the kind we're accustomed to in most years and seasons, is good. Long, withering dry periods punctuated by flooding downpours yield the same average rainfall, but not the same practical result.

"If you think of steps as leading up to a higher level, that's climate change. Climate variability is the height of each step," says the ecologist Steve McNulty, leader of the U.S. Forest Service's southern climate change research team. "The part that we know is changing is the intensity of the rain. That's a given, a very likely scenario in the climate models. With climate change, we may not see any difference at all in total precipitation. But with climate variability, that rain may all come in a six-month period, as torrential downpours, so we're likely to have more flooding."

The data on rainfall already seem to show a trend in that direction, according to one study of flooding: "Between 1948 and 2006, there appears to have been a 25 percent increase in the frequency of extreme precipitation events in Virginia. Extending the dataset to 2011, there appears to have been a 33 percent increase in the frequency of extreme precipitation events, with the 1-year storm now occurring every 9 months. As the frequency of extreme events has increased, so has the amount of rain that those storms produce (i.e. the biggest storms are getting bigger), with Virginia seeing an 11% increase in precipitation from the largest storms between 1948 and 2011."

Climatologists are trying to improve models for soil moisture that take the interactions of heat and precipitation into account. "Since all the models project warming, plants are going to need more water, and the relative effect of the warmer climate on plants will be drying," according to Ken Kunkel, a senior climate scientist at NOAA. "A second element is, when the models show drying, it is most prominent in the spring and summer, particularly in the Southeast. So just when you need the most water is when you are getting the biggest drying." Obviously, that would jolt the region's natural systems, and Virginia's agriculture.

It nearly goes without saying that a major portion of Virginia's food supply is produced somewhere else, and price and supply are dependent on the climate across those larger networks of sources. "Adverse impacts to crops and livestock over the next 100 years are expected," the most recent federal assessment for the U.S. states. "Over the next 25 years or so, the agriculture sector is projected to be relatively resilient, even though there will be increasing disruptions from extreme heat, drought, and heavy downpours," the report adds. "By mid-century, however, when temperature increases and precipitation extremes are further intensified, yields of major U.S. crops are expected to decline, threatening both U.S. and international food security."

One of the deeper strands of uncertainty woven into climate modeling is the underlying assumption that the historical patterns — the general rules governing what happens when climates warm or cool — will still hold. As climate scientists readily acknowledge, the "novel climate" we are creating, one that nearly defies comparison, may well invent some new rules.

Just as important as longer-term averages of rain, drought, and heat is this climate variability: how and when weather comes. Some climate models show occasional bouts of extreme cold at some times and places, compared with historical averages. Wags suggest that the term "global warming" be abandoned, at least for regional areas, in favor of "global weirding."

Daily readings of atmospheric CO_2 concentrations now exceed 400 parts per million. The atmosphere has not included this much CO_2 since about

3 million years ago, when the climate was a lot wetter and hotter. CO_2 is accumulating hundreds of times faster than it did as the last ice age ended. Or, by some other estimates, the rise in CO_2 is steeper than at any time since 56 million years ago, and perhaps longer — about the time, coincidentally and just for perspective, that our first known primate ancestor, a one-ounce, arboreal insect-eater, appeared on the planet.

THE CLOCK OF THE BAY

Not that it attracts much attention here in a picnic park along the suburban shore of the Severn River, but the scene is more than a little askew. On a drowsy, late-June morning on this Chesapeake tributary, there's the expected casual line-up of kayaks and Zodiacs, a high-schooler mowing the grass, some pleasure craft bobbing at anchor.

But the picnic tables are laden with an assortment of notebooks and odd tools. Some purposeful college students have unloaded tall green pressure tanks and meters that could be mistaken for a welding kit, and installed them in a 5-foot plastic tub under some trees. From there, tubes run down into Round Bay and out to a 10-foot square of water demarcated by colorful floating foam noodles. They are attended by two students in snorkel gear.

The low-tech look of the apparatus belies the high-stakes questions behind the research. Designed as a kind of time machine, it is meant to simulate these waters as they will be in the future, circa the years 2050 and 2100. The big green tanks are filled with CO_2, the gas that we now concentrate in the atmosphere as we burn coal, oil, and forests across the planet.

Surplus CO_2 has immediate impacts here. It is altering the basic chemistry of the Chesapeake and the Atlantic offshore of Virginia, as well as all other estuaries and oceans on the planet. They are already 30 percent more acidic than they were 250 years ago, in the pre-industrial era, because of added CO_2. That acidity is increasing at a faster rate than at any time in the last 300 million years, according to the most recent overview of the research.

A report requested by Congress from the National Academy of Sciences presents the case starkly: If the ocean continues to acidify, the water could become corrosive to calcium carbonate structures, dissolving coral reefs and the shells of marine organisms. As the Smithsonian Environmental Research Center (SERC) marine ecologist Denise Breitburg says, "There's no reason to think the same wouldn't be true in Chesapeake Bay."

Are changes in Bay organisms already occurring due to acidification? "Nobody's really been looking," says the SERC marine ecologist Whitman Miller. If the study of ocean acidification is in its infancy, the research on estuaries like the Chesapeake is embryonic.

The Severn River project on the Maryland shore is a collaboration between Miller and the chemical ecologist Tom Arnold of Dickinson College. Arnold monitors what happens to a patch of widgeon grass growing under that square of water offshore — especially its protective compounds that ward off predators and diseases — as CO_2 levels are artificially manipulated by releasing gas from the tanks. Miller's work, and that of several other Bay researchers, has been to chart the effect on oysters as they mature over the summer, sitting in a basket amid plumes of elevated CO_2 and rising acidification.

Miller was also the lead author of earlier research — a lab experiment. It found that, as CO_2 increases from pre-industrial levels to what's expected in the Bay by the year 2100, the shell area of maturing native oysters decreases by 16 percent, and their calcium content by 42 percent.

The fear is that acidification will make it more difficult for the oysters to form their calcium carbonate shells. The process has been referred to as "death by dissolution." It has already been seen in the ocean, among calcifiers such as sea butterflies, some planktons, and corals, and in severe declines of farmed oysters along the north Pacific Coast. In the Chesapeake it could begin to dissolve plankton and some or all of the shells of crabs, oysters, and 173 other species of shellfish.

Eastern oysters — the species name is *Crassostrea virginica* — are a sentimental favorite, evoking memories of a time when they supported a thriving seafood industry and communities of watermen and their families that are part of Virginia's mythos. Some are happy enough avoiding these slimy entities and their craggy, cement-colored shells at dinner. If you're an oyster devotee, though, you probably know that their flavor varies with the degree of salinity in the part of the Bay they come from.

Just as connoisseurs delineate the relative merits of microbrews or chili peppers or fine cigars, discerning fans of oysters compare the "sweet" versus the "briny" ones and the grades between. Distinctly flavored varieties are harvested from the waters near Chincoteague on the Eastern Shore, for example, or from Topping on the lower Rappahannock River, or from Milford Haven, Tangier Island, or Witch Duck Bay.

Oysters were once so numerous that the tops of their nearly limitless reefs were dry at low tide. "The abundance of oysters is incredible," a visitor wrote in 1701. "There are whole banks of them so that the ships must avoid them. A sloop, which was to land us at Kingscreek, struck an oyster bed, where we had to wait about two hours for the tide. They (the oysters) surpass those in England by far in size, indeed, they are four times as large." Some were 13 inches long.

Oysters are filter-feeders, pumping water into and then out of themselves, and they once functioned as a colossal water clarification system. It has been estimated that all of the water in the Bay used to be filtered by the vast oyster population in a week's time. That population has now been reduced by more than 98 percent.

There's still hope, an ardent fan club, and strenuous efforts on behalf of a modest recovery in the future. But generations of overharvesting and habitat destruction by industrial fishing technology, suffocating silt from upstream construction and tilling practices, the pollution of human and farm-based fecal matter and chemical fertilizers, and diseases probably imported from other parts of the world have all taken an inexorable toll.

Partly because of their iconic status, and despite the fact that they are so diminished, oysters are the shellfish species most often studied in connection with acidification. It may be the final blow to the tiny remnant of the wild oysters' former abundance.

Under ideal conditions, CO_2, sunlight, and water combine to drive photosynthesis for submerged grasses — another essential support system for the health of the Bay habitat. Since the 1970s, seagrasses have suffered declines large enough to threaten the whole aquatic ecosystem here. Because CO_2 is essential for plant photosynthesis, more of it might be expected to improve the grasses' prospects.

Instead, starved for light by silt pollution from farming and real estate development, seagrasses are unlikely to be able to take advantage of added CO_2, Arnold says. By the end of the season, his research had established that the experimental rise in CO_2 and acidification was followed by a sharp reduction of the grasses' coatings of protective compounds. "I've been doing this for about twenty years now, and it's the largest change I've ever seen," he said. "Everything we see so far tells us that it's not going to be so great for the seagrasses after all."

THE CHEMISTRY OF coastal systems and estuaries is much more dynamic than the comparative equilibrium of the ocean. In the Chesapeake, salinity varies widely, as do depths, tides, oxygen, temperatures, and the fluctuating acidity of freshwater streaming in from the Bay's 64,000-square-mile watershed. "Those are very strong signals, but nobody really understands them very well," Miller said.

All that variation makes it harder to measure in any simple way how much the Bay has already acidified. The processes through which it happens are straightforward, however. In the great out-of-doors, we think of sky and water as distinctly separate, but, in fact, where they meet, they mingle. The increasing CO_2 gas pressure in the atmosphere seeks its own equilibrium, pressing on the surface of open water and steadily diffusing into it, especially when the water is disturbed by wind and waves. It forms carbonic acid.

Explaining the resulting acidity calls for a bit of high-school chemistry. The strength of acid and its chemical opposite, alkalinity, is measured on the pH scale, which runs from 0, the strongest acid, through 7, neutral, to 14, the most alkaline. So as pH falls, acidity rises. Distilled water has a neutral pH of 7; lemon juice, 4; vinegar, 3; stomach acid, 2; sulfuric acid, 1. "Full-strength" ocean water usually had a pH of around 8.2. It has now declined to around 8.1, and is heading toward the 7s.

When we hear numbers like those we usually think in terms of rulers, measuring cups, or bathroom scales. A rise or fall measured in tenths hardly seems alarming. But the pH scale is logarithmic. Each full unit of decrease in pH (from 8 to 7, for example) represents a tenfold increase — that is, a 1,000 percent increase — in acidity. So the decline of .1 on the pH scale already measured in the oceans represents a 30 percent increase in acidity. Values of 7.8–7.9 are expected by 2100 in the "business as usual" emissions scenario, representing a 100–150 percent increase in acidity.

To repeat: acidity in the ocean has already increased by about a third. Tom Arnold pointed out that just that much might diminish the number of calcium carbonate ions beyond a tipping point. "That little window is the difference between some organisms having a shell or having a skeleton and not," he says. (Acidification is lately called "the other CO_2 problem.") "If we have a shift in the baseline," he said, "that could mean bad news for lots and lots of organisms, no matter how adaptable they are."

The idea is worth dwelling on. You can think of the Chesapeake's, and the Atlantic's, normal cycles of acidity as rising and falling within certain limits, as if they were inside a picture frame. Aquatic life has adapted to live within the limits of those cycles, but survival rates fall off near the edges of the frame, where the acidity is comparatively high or low.

If you move the whole frame a little at a time — as atmospheric CO_2 rises, for example — its margins begin to cut into survival chances of more and more kinds of organisms. The Miller-Arnold experiments are moving the frame, simulating the acidification process that is an evolving threat to the already sorely challenged, unstable health of the Bay.

Because these are simulations of the years 2050 and 2100, they may give the illusion that we have plenty of time to figure things out and act. The changes are incremental, though — they won't arrive all at once, a few decades out. "Fifty to one hundred years is not a very relaxed schedule, actually, given the magnitude of the problem," Miller says. "We're fighting against a time clock here."

Most of the acidification occurring in the Bay now is driven not by atmospheric CO_2, but by the more familiar problem of nitrogen and phosphorus pollution from farm fertilizers, and from sewerage. These nutrients lead to explosive growth of phytoplankton that, like all plants, absorbs CO_2 and generates oxygen. When it dies and decays, however, it releases the CO_2 again. The upswings are much worse than the contribution from the CO_2 in the atmosphere, at least for the time being.

So those larger factors are already boosting acidification levels that test the tolerances of life in the Bay, but they are controllable as soon as we act. The additional acidification from climate change will advance, either way.

"We don't really know how the Bay is going to respond," the ocean ecologist George Waldbusser told me. Three of the conclusions in his research are striking, however. At current average pH levels in some parts of the Bay, the rates of shell growth are already slowed, so the shells of young oysters are abnormally thin, making them more vulnerable to predators. The saltier Bay waters are likely to become more acid, increasingly corrosive to oyster and other shells. They are already unsuitable for shell preservation in some areas that once supported healthy oyster populations.

"We've altered coastal ecosystems in ways that have affected that carbonate chemistry and already pushed the system to levels that were predicted for a hundred years from now, or even further into the future," Waldbusser says. "I think we're worse off than the open ocean." Scientists who research acidification in the Atlantic, however, are anything but relaxed about the prospect of climate change there.

VIRGINIA'S OCEAN ESTATE

The research vessel *Ronald H. Brown* is rolling a little, out ahead on the waves an hour east of Norfolk. It's a fogged-in, late-spring morning, and at least one of us is a little seasick as we approach on our small, plunging commuter boat. We're here to retrieve about a dozen scientists who have mapped the ocean floor for the past three weeks and studied a group of mysterious and little-known animals, some of them a mile below the surface.

Virginia's economic claim on offshore resources—oil- and gas-drilling revenues, for example—extends 3 miles from land. It is shared with the federal government for another 3 miles out, where exclusive federal jurisdiction begins. By virtue of our membership in the federal Union and thanks to international law, however, Virginia also enjoys some of the wealth of what's called the Exclusive Economic Zone. That reaches 200 miles from the coast.

The scientists on the *Ronald H. Brown* have been reconnoitering a kind of wealth that we know little about and whose value we cannot readily guess for now, even with their help. The surface of this part of Virginia is familiar enough, at least to fishers, but its depths are for the most part *mare incognitum*. For example, despite a marked absence of tropical beaches and warm, sunny shallows, there are at least a dozen species of corals out here—vivid spikes, balls, fans, and trees—a key part of ecosystems that until recently have been inaccessible, too far down to study.

These deep-water, cold-water corals occur mostly along the sheer walls of miles-long rents in the ocean floor, at the edge of the continental shelf. They are canyons, invisibly remote but shaped like zags of ancient lightning on the new sonar maps the research project has drawn for us.

Norfolk and Washington canyons crack the seabed open some 60 miles off the coast. Their areal extent taken together — just the portions that are landward of the break of the shelf — is about 58 square miles, big enough to swallow the city of Richmond. The canyons also run miles seaward on the ocean floor beyond the shelf, and to a depth of about 6,000 feet. They are the southernmost, and among the largest, of a series of such incisions that run up along the east coast as far north as Maine.

"I am really surprised at the abundance of bubblegum coral down there — it actually does look like bubblegum — and there are massive trees of it," Sandra Brooke, a deep-coral specialist, told me. "One was about 15 feet tall." The species forms an elegant, electric-pink candelabrum on the video sent up to the mother ship by *Jason,* an unmanned submarine used for the research.

The researchers also came across the unexpected, ghostly presence of a stony coral called *Lophelia* while exploring canyon crevices during the project. The species was thought not to occur at all along a mid-Atlantic gap of several hundred miles. "We've had very little information about the communities in the canyons," Brooke said. "The rockier habitats in these obscure places are just a big unknown. We've never been able to look there before with the kinds of tools that we need."

The research vessels were provided by NOAA, and most of the funding by the Bureau of Ocean Energy Management (BOEM). One of BOEM's keen interests: reckoning the potential environmental impacts if it grants permits for future oil and gas drilling, or wind-energy platforms. "We want to know what's down there so we'll know what to avoid damaging," a spokesman told me.

The scientists, however, also want their research to help make the case for the establishment of new marine sanctuaries. They hope to protect an offshore treasure that is already mauled by trawling, overfishing, and pollution — and that may face an even more comprehensive and enduring threat, acidification.

Virginia has established "no take" sanctuaries — meaning no fishing, drilling, or other incursions — on a minuscule portion of the Atlantic within its jurisdiction: about .02 percent. That's one-half of 1 square mile out of a total of 2,767 square miles of marine area under state control. Nearly a quarter of Hawaii's waters are in no-take sanctuaries; nearly 10 percent of California's. Several other coastal states outrank Virginia by huge multiples.

OUT HERE, the edge of the Continental Shelf formed the Virginia shoreline nineteen thousand years ago. The last ice age was at its maximum, and some of the water of oceans and rivers was locked up in glaciers and polar ice sheets. Sea level was about 360 feet lower as a result. The high-and-dry rims of the big canyons back then formed inlets, like fjords, along the coast. They were the farthest extensions of down-cutting rivers that have been submerged, by the processes of deglaciation and sea level rise, for millennia.

Reconstructing the ebb and push of these long cycles of sea level rise and retraction, warming and cooling, is the portfolio of a relatively new branch of science called paleo-oceanography. Lately, its practitioners also chart other oceanic rhythms in deep time. They analyze the chemical signatures of fossil material collected from the ocean floor, which can tell us about variations in ocean temperatures and acidity over the eons, and their consequences for life on an evolutionary scale.

That is the agenda of a Columbia University lab run by the paleo-oceanographer Bärbel Hönisch. "What people want to know today is what's happening to the fish, to the oysters, to all kinds of other organisms," she told me. "But very often we have to look for those plankton fossils, like the *foraminifera,* that leave a widespread record."

Hönisch led a team that put together an overview, recently published in the premier journal *Science,* of what we know so far about earlier bouts of ocean acidification on the planet. "One of the reasons why we wanted to look at the paleo record is that we can compare it with what is happening today and improve predictions for future changes," she said.

As noted in the previous chapter, acidity is measured in log-scale pH units, and as pH falls, acidity intensifies. The 0.1 unit of decrease in "full-strength" ocean-water pH from around 8.2 to 8.1 means that a 30 percent increase in acidity has occurred. Given the current rate of CO_2 emissions, Hönisch said, "we are expecting another decrease of 0.1 to 0.2 in ocean pH by the end of the century, or maybe even .3."

The amount of acidification is crucial, but so is the speed at which it occurs. The waning of the last ice age also resulted in an era of marked acidification. The difference is that we're now pushing CO_2 into the atmosphere at a rate hundreds of times faster. As noted earlier, ocean acidification may now be progressing more quickly than at any time in the last 300 million years — which is about as far back as the record of seafloor fossil sediments allows us to look.

The Hönisch study concludes that the current acidification rate raises the possibility that "we are entering an unknown territory of marine ecosystem change." There is nothing comparable to it in the known geological record. In earlier periods, at least some organisms may have had time to adapt to these conditions, she told me. Those acidification periods took place over thousands of years or longer, however.

The cause of one such occurrence around 56 million years ago, the Paleocene-Eocene Thermal Maximum (PETM), is uncertain, but it left strong evidence of ocean acidification and warming. "That is probably the event that looks most similar to what we're doing today, so that would probably be the best analogue," Hönisch said. There is plenty of evidence of extinctions among a wide range of organisms during the PETM. Fifty percent of the bottom-dwelling *foraminifera* plankton species vanished, for example.

"What is important about them is that they are very simple organisms at the bottom of the food chain," Hönisch said. "What you can infer from that is that if they became extinct, then their disappearance would have propagated higher up in the food chain. Higher organisms would have been affected as well." We are already seeing the impact of the acidification and warming we have now, she pointed out: a decline of 14 percent in the growth rate of corals on Australia's Great Barrier Reef, for example.

Underwater volcanism can cause localized acidification of seawater in a few spots around the world. In these places, where there would be healthy, diverse ecosystems if pH were normal, most of the shell-making organisms are missing, or their shells are damaged.

I posed the all-too-obvious question toward the end of our conversation: given the changes we are seeing in the ocean, are we at the edge of another mass extinction event now? "That is really hard to predict," Hönisch said. "It might happen or it might not. In my view, if you can't predict it, it makes sense to be careful. It's like driving in the dark without your headlights on. It makes sense to slow down, not to accelerate, or by the time something happens it is too late to do anything about it. I think the likelihood that something will happen is rather high.

"What are we going to do? I guess we can convince our politicians that we have to spend a lot of money on these things. We can convince ourselves that we have to tighten our belt more and not be as exuberant in everything that we want to do. I think we have to start with everybody. We can't smell or see the CO_2. If you go into the ocean you can't feel that it is more acidic. It's hard to convince people to take this seriously."

A SLOW, GRINDING synergy of warming, acidification, and deoxygenation may already be showing up. Along the Mid-Atlantic Bight — the region east of Virginia and the long coastal curve from Cape Hatteras to Massachusetts — sea surface temperatures during the first half of 2012 were the highest ever recorded, according to NOAA's Northeast Fisheries Science Center (though the record is not long).

"Above-average temperatures were found in all parts of the ecosystem, from the ocean bottom to the sea surface and across the region . . . and the spring plankton bloom was intense, started earlier and lasted longer than average," the agency's advisory said, adding that this affects a wide range of marine life, from plankton to whales.

"A pronounced warming event occurred . . . and this will have a profound impact throughout the ecosystem," the NOAA scientist Kevin Friedland warned. "Changes in ocean temperatures and the timing of the spring plankton bloom could affect the biological clocks of many marine species, which spawn at specific times of the year based on environmental cues like water temperature."

A natural pattern of ocean warming and cooling called the Atlantic Multi-Decadal Oscillation and a lack of long-term historical data about these temperatures fuzz the picture, but several recent studies have confirmed significant warming in the world's oceans.

Meanwhile, fish are responding. The movements of thirty-six fish populations from Hatteras to Canada were charted in a study led by the fisheries ecologist Janet Nye. Those populations, many of them commercially valuable species, have been shifting north and eastward over the last four decades, and some have all but disappeared from U.S. waters as they move farther offshore.

Atlantic cod and haddock, several kinds of hake and flounder, alewife, blackbelly rosefish, a big relative of the cod called cusk, goosefish, and other species have been migrating to cooler waters. Several of those species along with Acadian redfish and wolffish are also moving into deeper waters than where they have traditionally been found: "They all seem to be adapting to changing temperatures and finding places where their chances of survival as a population are greater," Nye told me.

She added that yellowtail flounder and other species that typically do not migrate cannot readily adapt by moving long distances. "Their populations will probably decline, and for fisheries that is a big deal," she said. One of her chief concerns is the combined effects of overfishing and climate: "Yellowtail flounder is more of a northern fish, and we used to have a really healthy population off the Northeast coast. They haven't recovered in two decades. I suspect there are a lot of fish populations that just haven't recovered because of climate change."

Friedland and Nye were part of a study that documents the decline of plankton because of increasing ocean temperatures in the Northeast — the same possibility that Hönisch worries about. The change puts further pressure on dangerously overfished cod populations. Fish can tolerate some degree of acidification, but they depend, one way or another, on the plankton at the bottom of the food chain. "There may be a profound change in the productivity of the ecosystem at the bottom — what's going to happen to the plankton?" Nye wonders. "The fact that mass extinctions occurred in the past when CO_2 was increasing at a much slower rate than it is today is definitely worrisome."

Many formerly abundant fish populations are now small and isolated, fragmented by generations of overfishing. "When you look at maps of Atlantic cod, and they've been fished very heavily, you just see little tiny hotspots of them," Nye said. "In order for fish to shift their distribution and colonize new areas they have to have large enough populations. To see that a lot of these populations are already so fragmented that they might not be able to adapt by moving is worrisome. I think some species will be in trouble."

DURING THE coral-and-canyons research cruises at what used to be the edge of ancient Virginia, seeps of methane gas have also been discovered on the sea floor. Unrelated to climate change, as far as is known, they may be the result of slowly decaying buried organic material from that long-ago epoch when the shelf was all dry land. The seeps may turn out to be far more common than once thought, the University of North Carolina fisheries biologist Steve Ross said.

The cruise retrieved mussels from one seep in Baltimore Canyon whose nearest populations are found in the Gulf of Mexico — an example of the odd and displaced biology that is supported by the upwelling methane. "One of the most interesting things about these seep communities is they are chemosynthetic rather than photosynthetic," Ross said. "It's the only life-sustaining activity we know of that is not based on energy from the sun. It shows that sunlight is not necessarily essential to some life forms, and that has huge implications for looking for life on other planets." A sardonic silver lining: if global warming or some other planetary hazard interferes with photosynthesis, he speculated, "these systems will probably just keep right on going."

Ross describes Norfolk and Washington canyons as funnels for strong currents that sluice in crustaceans, plankton, and other nutrients to create a rich feeding area that supports a dense array of sea life. They are heavily packed with squid and with blackbelly rosefish, pilot whales, sperm whales, caves full of American lobster and golden tilefish, and the corals.

The vandalism from overhead is also much in evidence. "Trawling is probably one of the most destructive fishing practices in the world," Ross said. "Imagine a field of six- to nine-foot corals just mowed down by trawl nets. We saw so much damage and trash, plastic bags wrapped around corals, garbage, discarded nets, traps, and lines. I've never seen that much junk on good habitat in thirty-plus years of doing this kind of work. And a lot of the deep-sea fish that are caught by trawling are old and slow-growing, and they don't reproduce much, so they're wiped out easily."

Some of the research that Brooke and Ross have led on the *Ronald H. Brown* is to set up monitoring of ocean acidification. Although they are stationary like plants and look like them, the deep-sea corals are animals, with skeletons and metabolisms. Though continuing acidification is a given, it is not yet clear what its consequences will be in the deepest waters, Ross said.

Brooke noted the irony: the cruise project's discoveries are being made just as these grim synergies gather. "We don't know what most of those fish are eating. We don't know where their juveniles are. We don't know if any part of their ecology depends on the coral. We're still only at the point where we are inventorying these animals, let alone knowing what their role is to commercial species or to each other," she said.

What may we take for granted here? Virginia is the third-largest marine producer in the United States, with total landings of about 250,000 tons a year. The dockside value to watermen alone is close to $200 million a year — the largest seafood production of any state on the East Coast. Hampton Roads is the seventh-wealthiest seafood port in the nation. As is the case nearly everywhere, pollution, destructive fishing practices, and overfishing have ill-used the Bay for generations, and have depleted our part of the Atlantic, too. This will all make ocean ecosystems less resilient under the stresses of warming.

Oceans serve as the world's largest source of protein: more than 2.6 billion people depend on them as their primary source of it. Marine fisheries directly or indirectly employ more than 200 million. Overfishing has already resulted in 90 percent of the world's fisheries being "fully exploited, over-exploited, or collapsed," according to the most recent research overview.

A recent National Academy of Sciences report anticipates that climate disruption may kill off 60 to 100 percent of all coral reefs before the end of this century. They, in turn, support between 9 and 12 percent of the world's total fisheries.

"We are busy destroying things when we really don't know their value," Brooke said of Virginia's fraction of this general picture. "And when we do — well by then, you know, the horse has not only bolted. It's got rigor mortis."

TAPESTRY, INTERRUPTED

The trail to the top of Mount Rogers meanders over a couple of jutting rock ridges, then crosses a grassy bald and a few patches of conifers. On some days it also leads past shaggy wild ponies, so often admired by hikers that they're pouty, though they will condescend to eat an apple if you have one. Then the trail climbs into broken shade under a wind-battered canopy of red spruce and Fraser fir. It's often foggy and frigid here. Rain and snow are frequent. At 5,729 feet, the summit is the highest point in Virginia, and the mountain is a refuge from global warming—the warming that got under way at the end of the last ice age. Soon it will be a trap.

I'm in boreal forest, named for Boreas, the Greek god of the north wind, and around me are plants and animals usually found in parts of Canada. It's an isolated remnant, a single square mile from that far colder time when the spruce-and-fir forest covered 700,000 square miles of eastern North America, including most of Virginia.

Cold-adapted forests like those, and the plants and animals that lived within them, migrated north as the barren edge of the Laurentide Ice Sheet, which once extended as far as Pennsylvania, melted back. Or they shifted upslope to escape the gathering warmth, into the colder zones of a few of the highest peaks in the Southern Appalachians such as Mount Rogers. Meanwhile, their southern margins slowly succumbed, outcompeted by warmth-loving species.

Each species moved north at a different rate, so the various ecosystems that had evolved during ice-age Virginia disorchestrated. They eventually formed new patterns with new players in a comparatively stable climate that has persisted for the past eighty centuries or so, until now.

That stability has resulted in the state's complex and celebrated tapestry of biological diversity, from cove forests to pocosins, Table Mountain pine outcrops, pine-oak heath, shale and limestone barrens, dune grasslands, tupelo and cypress swamps, oak-hickory forests, and Piedmont prairies, to name a very few. They all are home to 61 kinds of reptiles, 85 mammals, nearly 400 birds, more than 3,000 flowering plants, ferns, and shrubs, 238 species of trees, and 210 of fish.

Mount Rogers is the last remaining spruce-fir habitat in Virginia—actually the only place in the state where Fraser fir can still be found at all in the wild, along with a dwindling assemblage of other rarities such as endangered northern flying squirrels, three-toothed cinquefoil, Weller's salamanders, and nesting northern saw-whet owls not much larger than your fist.

Specially adapted to their patch of relict climate, these plants and animals are hemmed in. There's nowhere left to go to escape the encroaching heat. No physical corridors to colder places in the North, and no farther upward moves to be made, either. They're already here at the top of the highest mountain, the end of things. (Unlike people or animals, plants cannot relocate, but distributions of their populations can move as time passes. When I refer to "trees migrating," that's shorthand.)

Standing in this silent cathedral forest on a carpet of sorrel and wood fern brings a kind of eerie, involuntary double vision. You see what's here now, and then the drier, hotter, more biologically impoverished place that is all too likely to be its future. Average summer temperatures that are 5 degrees higher can erase this forest and the constellation of species that depend on it. That is within the range of "business as usual" projections of warming for this area of the state after midcentury. And that's just the average. Prolonged hot spells above the average are an even sharper threat.

Virginia's Department of Game and Inland Fisheries has a wildlife action plan coordinator, Chris Burkett, part of whose job description is to protect rare species and keep the others from becoming rare or extinct. That includes trying to think through the implications of climate change. It's a "mission impossible" in most ways, for him and for his counterpart planners at other state agencies.

Article XI, Section 1, of the Virginia constitution promises that "it shall be the Commonwealth's policy to protect its atmosphere, lands and waters from pollution, impairment, or destruction, for the benefit, enjoyment and general welfare of the people of the commonwealth." There's little in the way of research, staff, or especially political support and money to plan for the effects of climate disruption on natural areas, however. In reality, nearly any progress within state agencies on the climate change front is a remarkable achievement given the political context within which they work.

"I suspect we will see some dramatic changes on Virginia's landscapes, and lose some species," Burkett predicts. There is no plan to try to alter the fate of the spruce-fir forest and its wildlife on Mount Rogers, or for the scatter of other, similar forests on nearby peaks. "To be honest, I'm not sure anything can be done to conserve them," he said. "I don't see a way that we could do much in Virginia for that habitat."

Though Mount Rogers's "sky island" ecosystem is small and unique, a similar predicament will become increasingly common on the 15 million acres of other Virginia forests, which blanket more than 60 percent of the state's landscape. As the climate heats up and many plants and animals have to migrate northward or upslope if they can — or die off — these questions will press harder: What can replace them, and how quickly?

There's a sharp difference, as noted earlier, between the current warming and the earlier passage of the planet into a warmer epoch. The projected pace of change now is much faster than the exit from the last ice age. The climate record shows that the state's mountain regions are already registering higher temperatures — a steady upward trend in all seasons over the past four decades.

BURKETT'S AGENCY has a sketch-plan and modeling data for some bellwether plant and animal species — a preliminary conversation about protecting the rarest ones from extinction in the face of climate change. There aren't just a few rare and threatened species in Virginia. There are more than nine hundred.

Only half a lifetime ago, conservation biologists were preoccupied with single species that might become extinct. The catch-phrases of the time reflected an outlook that seems optimistic now. Individual species in trouble were said to be early warning signals — like canaries in coal mines — or small but essential, like rivets in airplanes. They were the works and windings of natural systems, which no sane tinkerer would cast aside.

But it quickly became apparent that in order to save a checkerspot butterfly, a dusky seaside sparrow, a grizzly bear, or some other species, you have to figure out how to protect a large, sustaining portion of its habitat. It must include the other species essential to the survival of that community, and more and more of those were also discovered to be at risk. The habitat has to provide a buffer against pollution, hunting, logging, roading, and construction, forever. A functioning Noah's Ark was going to need more than just the plants and animals, in order to bear the growing lists of the vulnerable into an uncertain future.

Then it began to register that not just a few species or habitats were in jeopardy, but whole classes of animals across a wide geography. So many that the politically impotent U.S. Fish and Wildlife Service fell years behind in trying to process candidates for its national lists of the "threatened" or "endangered" that were supposed to be under federal protection. In Virginia, 46 percent of our fish species, 25 percent of birds, 46 percent of reptiles, 43 percent of amphibians, and 28 percent of our mammals, for example, are now considered to be in those two categories, or shrinking in population.

At this point, many of the causes of the mayhem have long since come to be recognized as regional or continental. The acid rain sterilizing soils and streams in Shenandoah National Park and on Virginia's national forests is largely the result of coal smoke that billows up and drifts in from power plants in the Tennessee and Ohio valleys. Ground-level ozone pollution also reaches here from tailpipes and smokestacks that are sometimes hundreds of miles away. So the conversation has moved from single kinds of frogs or flowers all the way to planetwide trouble: polar ozone holes and greenhouse gases.

You can be a "lumper" or a "splitter" in your analysis of different causes of environmental damage. That is, you might lump some or all of them together for purposes of assessing trends, or the net effects of our environmental misbehaviors. Or you might split up the list, to see which factors are the most destructive, the most costly, or the easiest to fix.

Global warming, however, is distinctive in its causes and its scale. It is the most far-reaching threat to the natural environment in its own right. But it also speeds up and synergizes environmental damage from many other causes such as air and water pollution or habitat destruction. For the forests, the farms, the Chesapeake, the James River, or the Atlantic coast, this means that while we try to cope with climate impacts, we will have to fight much harder along several other environmental fire lines, too.

Here's how the state Department of Game and Inland Fisheries' climate strategy document summarizes that outlook: "Over 900 of Virginia's wildlife species are believed to be imperiled by the ongoing loss or degradation of their habitats. During the coming decades, climate change will exacerbate and intensify these impacts and the consequences to wildlife could be profound."

For example, cold-water species like brook trout, which struggle when water temperatures rise above 70 degrees Fahrenheit, could lose much of their current range. "Unfortunately," the document says, "populations that cannot move to higher elevations, or fail to find suitable habitats, will be extirpated." The definition of that word is "rooted out and destroyed completely" in a given area.

AT THIS CHEERLESS POINT in conversations about climate change, a reassuring statement will sometimes be heard: "When things change, there are winners and losers." The statement can be flat wrong in a practical sense, of course. Burn down your home and then try to calculate wins and losses. But the generalization can indeed be true on a warming planet, on some broad biological scale. If we gaze out over a long enough period — say, a few million years — evolution guarantees that some living things will adapt, some new ones will arrive, and some will fade from the scene.

Even in the near term, at least some plants and animals will prosper. The prospects are especially bright for the hundreds of alien invasive species — the polite name is "introductions." They are usually from Asia and Europe, and we have continued to import them at an accelerating rate over the past couple of centuries with unreckoning enthusiasm: Japanese stilt grass, garlic mustard, kudzu, Chinese mitten, mile-a-minute weed, *Ailanthus* and *Paulownia* trees. Stinkbugs, fire ants, feral pigs, "starry sky" and lots of other beetles, borers, gypsy moths, and fungal blights. The whole long list of thriving invasives has frayed natural systems and all but banished chestnut, butternut, fir, ash, elm, and hemlock trees from Virginia's native forests.

Invasives typically specialize in disturbed landscapes. Often they can disperse quickly, tolerate a broad range of climates, and find it easier to colonize forests where their native competitors have weakened or disappeared because of acid rain, ozone pollution, erosion, intensive logging, and gas drilling. The torn-up ecology of climate change will enhance their opportunities.

The work of Kenneth Landgraf, the supervising planner for Virginia's George Washington National Forest (GW), makes it evident that the scale of climate change is far greater than the research, personnel, and funding currently available to do much about its impacts. Usually his job consists of creating plans for the management of recreational sites, for logging or prescribed burning, or for protecting rare plants and animals. "We had a lot of concern about climate change — okay, so we bring it on as an issue, but what are we really going to do with it?" he asks.

A Forest Service catalogue of a couple of hundred climate disruptions anticipated for the GW makes plain the mismatch between the threat and the potential responses to it. Each expected change from the heat, the drought, the invasive species, or the fires was followed by a to-do list. Dozens of impacts carried only this brief notation for management options, however: "none identified."

Possible actions were indeed listed for many of the other threats. But the draft management plan for this national forest, intended to be in effect for a decade or so, includes only a handful of worthwhile but poignantly small, passive steps. They cast Forest Service managers, with their limited resources, in the role of bystanders at a train wreck. More effective action to protect these public lands and resources will require a far stronger federal commitment, and major changes in how they are managed.

The draft plan for the GW merely identifies some species vulnerable to climate change and expresses the need for more monitoring and for more control of invasives without specifying how or when these initiatives will occur, how long they will continue, and how they will be funded. Some culverts, altered so that brook trout might escape to cooler streams as their habitat heats up, are recommended. Landgraf said the possibility that even that small measure would succeed may be "remote."

"What we were trying to do is acknowledge the fact of climate change and get it into our plans, so that we continue to at least think about it as we move forward. But knowing exactly what to do with the information is still the hard part," he said.

Ailanthus — often called "stinky trees" in their native China — and dozens of other imported species make it plain that controlling invasives in Virginia's forests is already all but impossible, given our current levels of inattention. Landgraf recalled a recent excursion to a remote part of the national forest that was being considered for formal designation as a wilderness area. The group came upon a section that hadn't been logged in forty or fifty years. They found an unwelcome surprise there in the hoped-for wilderness: clumps of *Ailanthus*.

"Normally, they're near open areas or roads, but even deep in the forest, we're finding it," he said. "*Ailanthus* can come in and totally dominate the stand at the expense of most everything else in there. So that's one that's getting pretty scary and unfortunately, more and more extensive." *Ailanthus* isn't in the woods because of climate change, but it is poised to take full advantage of it. Invasive plants, insects, and diseases, several research studies have concluded, are highly likely winners as warming advances.

They cannot substitute for our existing native forests in supplying what scientists and economists call "ecosystem services," though. The term is an attempt to account for the value of natural systems to humankind — for our agriculture, standard of living, water supply, air quality, the economy, and, sometimes, for our survival. One partial estimate pegged the economic value of the ecosystem services produced by Virginia's forests at nearly $12 billion each year.

They are not an intense part of daily experience for most of us, but forests are more than just scenery. Their shade cools soils and streams. They provide wood fiber and lumber, moderate runoffs that can lead to flooding, and hold soils against erosion, especially in the mountains — one of the principal reasons the national forests were established. Their filtration processes help protect air and water quality.

Trees also absorb billions of tons of the planet's CO_2, storing it until they decompose or burn and release it back into the atmosphere. If you drive from Norfolk to Staunton, or on west to Abingdon, you're often traveling through tracts of young forest, regrowing on what used to be small family farms, and they're in fast-growth, fast-carbon-storage mode. Old trees also pack on wood growth and carbon storage quickly, growing in diameter rather than height, recent research has found. In many ways, the older the forests, the better, and the less logging and clearing, the better.

When you look at a tree, you are seeing an organism that's designed to capture radiation coming in from the sun — it traps heat and controls it by evaporating water. Landscapes with trees, then, have the advantages of a heat-capturing, water-emitting blanket that cools the local climate with shade as it locks up carbon, at least for a while.

I WALKED WITH THE forest ecologist Hank Shugart on a trail through the 200 acres of oak, ash, pine, hickory, and poplar at the Montpelier estate. It's a long way from Mount Rogers, and much more representative of the mix of trees across most of the state — similar to the poplars, pines, white oak, pawpaw, red maple, and red cedar in my own yard. Montpelier is also quite rare, though, in a different way. This is an "old-growth" forest, more or less unaltered by people. It was President James Madison's homeplace.

No one quite knows why this hillside in rural Orange County seems never to have been logged. It is prime farmland, platted in 1723 along the same low ridge as Thomas Jefferson's Monticello, 30 miles to the south. But Montpelier historians say that in his speeches and writings President Madison worried about the loss of Virginia's intact forests, increasingly plain even in his time. The oldest tree found here so far is an expired white oak that dates back to 1712 — twenty years before George Washington was born.

Shugart had been studying forest dynamics for thirty-five years at the University of Virginia when we talked, and he's on intimate terms with these woods. He said there's evidence here, in Madison's own meticulous climate records and in the rings of the old trees, that summer rainfall patterns have shifted. Other data confirm that spring arrives earlier now, and the growing season is longer.

Shugart's statistical models of forests such as this one, when they're driven by a continuously warming climate, echo the scenarios projected in several other such research efforts. They show not much change during the early decades. Then the response can be abrupt: a massive dieback from heat stress, fire, insects, or drought, brought on by the warming. After that comes a gap of indefinite duration before new sorts of trees can migrate in.

That conclusion was part of his testimony at one of the public hearings of a short-lived Virginia state commission on climate change during the Timothy Kaine administration. Looking out at the foothills around Montpelier, Shugart—a self-described optimist—recalled that preparing his climate commission remarks evoked some of the dread that many scientists say they feel at times.

"One of my thoughts was to try to put a pleasant spin on anything I could," he said. "For me, giving that talk was unpleasant, because I actually had to think about that stuff in ways I normally don't. I usually detach myself from it. So it's funny, when I put this little talk together, one of the things that ran through my head was, 'God, this is kind of bad. It's really . . . it isn't good.'"

The beginnings of those changes have also been forecast by Dominique Bachelet, a senior climate change scientist at the Conservation Biology Institute in Oregon, using forest ecosystem models that incorporate a long list of factors, from humidity to the heights of various kinds of trees. They produce a variety of results, but all point to marked change in southeastern forests.

Forecasting precipitation is intractably uncertain in the Virginia region, she noted, echoing every other climate modeler I've asked. But she said it is highly likely that advancing heat will dry out the landscape even if rainfall increases, because of evaporation. According to the latest federal assessment for the Southeast, that will increase wildfire risk across the region's forests. Because of the dry-out, the study says, "continued increases in metropolitan populations coupled with increased water use by forests will likely cause more frequent and severe regional and local water shortages."

Examples of the scale of the potential impact of temperature increases on forests aren't hard to find outside Virginia now. Tens of thousands of square miles of forests across the Mountain West and Canada are dead or dying. Intense droughts and hotter temperatures have generated both wide-scale insect infestations and fires — perhaps the worst since Europeans first arrived.

As fires become more frequent, savannahs with only occasional trees could replace Virginia's forests. "Savannafication of the Southeast . . . could be one of the most profound potential climate change impacts in the U.S.," a federal study says.

As heat advances and moisture recedes, the rhythm of fires may quicken again, transforming savannahs into grasslands with no trees at all. Bachelet has concluded that "the climate models have actually been relatively conservative. . . . It's much easier to see climate change as linear, but in fact, that's not what's happening. We could have some periods that are much more severe than the climate models are simulating. So everything is possible at this point."

Her coauthor, the bioclimatologist Ron Neilson, adds that some models predict large areas of forest in Virginia and the rest of the Southeast that "go into drought stress and potentially burn up." In the hotter scenarios beyond 2075, "the amount of fire that's showing up is pretty horrific," he said, resulting in much more extensive grass and shrublands and many fewer trees.

A preview of those scenes flared in Virginia and southwards, especially in late 2016, when more than one-fifth of the whole Southeast was suffering "extreme" or "exceptional" drought. Dozens of wildfires burned for weeks, charring more than 80,000 acres across the North Carolina and Georgia mountains. Smoke choked distant cities, spiking hospital admissions for respiratory distress. "Everywhere smells like a campfire," one public health official said.

"My main mantra, frankly one of the things I'm very concerned about, is the potential for forests to turn into a conflagration, particularly in the Southeast," Neilson told me. In the relatively near term? "Absolutely." Twenty or thirty years? "How about now?" he replied.

POND AND PARADOX

Animated maps show the slow-motion metamorphosis of the forest on the long reaches of eastern North America as the grip of the last ice age weakened. You can dial in a tree species and then watch as the millennia flip past, the blankness of the continental ice sheet retreats, and clouds of fir, beech, oak, or pine pulse northward, responding to the pressure of a warming climate. The maps are based on the faintest of traces in the stratified mud of remote ponds and bogs.

In Virginia, one of these troves of ancient natural history is hidden within a mountain forest in Allegheny County, a long way from pavement. It is a tiny natural pond of less than an acre, rare in Virginia, on the flank of a mountain called Tom's Knob. Silt has sifted into it, layer on layer over the centuries since it was formed some 13,250 years ago.

Paleovirginians shared that ice-age landscape with *Megalonyx jeffersonii*, the 800-pound ground sloth, as well as with eight-foot-long beavers, musk oxen, mammoths, mastodons, and the giant short-faced bear. Fossil evidence for that whole menagerie has been exhumed at what is now Saltville, a few counties distant from the pond — a dig that, when it began, won the support of our eminent natural historian Thomas Jefferson.

My wife and I hiked in to the vicinity one October afternoon over a steep and bouldery track, but we couldn't find the pond. Screened by a fringe of trees, it finally emerged — a bright mirror of rolling clouds, dotted with the stems of a nearly extinct bullrush.

Four decades ago, the Irish paleobotanist W. A. Watts found his way to this same lonely glade in pursuit of considerably smaller quarry: microscopic fossilized pollen. Its patterns in the pond mud signal what happened to the forests as the ice receded and spruce and fir began to give way to passing waves of other kinds of trees. The evolving climate led to the current mix of oak, hickory, maple, ash, and buckeye that filled the near horizon in a wild blaze of autumn color on this day.

Watts's research techniques probably did not differ much from how the work is done now. You lash a couple of canoes under a platform about the size of a picnic table, with a hole in the middle. Float that out to the deepest place, then ram a metal tube through the hole and down farther and farther into the mud, and deeper in time, to gather as much as 90 feet of a core sample. Back at the lab the layers are parsed and dated, using the decay of carbon 14 isotopes — radiocarbon dating — as the calendar stone.

Then the material is meticulously combed for bits of charcoal, plant and insect parts, and, especially, pollen, which reveals the relative abundance of different kinds of trees that were on the landscape, and how the mix changed over time. Pollen is gleaned from the mud using hydrofluoric acid strong enough to etch glass and dissolve sand — an indication of the durability that has preserved the tiny pollen grains through the passage of so many dozens of centuries.

The point of Watts's labor was to reconstruct the long arboreal pageant, but his kind of research has become rather more urgent now. Those past changes are being used to help us predict what will happen to the forest in the future as our climate movie careens into fast-forward. How quickly can tree species move, and in what manner will the forests endure? We look back to the postglacial period because that's the last time that tree populations have had to disperse over long distances in relatively short periods of time.

The maps of tree migrations make their movements look steady. Now, however, the broad reach of our activities has altered the life-chances of forests during the coming climate change in problematic ways. We are entering what some have called a "no-analogue climate." That is to say, we can't see the kind of abrupt change we're now facing in the known record of pollen, ice cores, or fossil tree rings. The climate is shifting so quickly that it could outstrip the pace at which plant species, including trees, can move. This will be a race between the migration of plants and animals on the northern edge of their ranges and their disappearance on the southern edge.

"It's kind of interesting," the forest ecologist Hank Shugart says. "We're looking at climate change in the future, perhaps occurring in a century or less, that's about equivalent to the transition between the last ice age and now, which just absolutely rearranged the planet." And, of course, Virginia. That change took place over thousands of years, though, not just a hundred.

Also, for the first time in the planet's history, our farms, highways, urban and suburban expanses of homes, shopping malls and office parks will form a biologically sterile barrier to plants and animals as they try to migrate in response to the climate, and it will halt many of them.

Scientists once thought that tree populations could spread in response to climate change, generally speaking, at a rate of about 300 feet to a half mile or so each year, depending on the species. The high end of that range is fast — 500 miles in a thousand years. Fast enough to have kept up with the push of post-ice-age warming. But more recent research has concluded that even under optimal conditions, the rate that the trees can disperse is really more like only a half mile each five or ten years. The high end of this calculus is only 100 miles of migration in a thousand years.

So the climate may change a hundred times faster than during the last big climate shift, but the rate that trees can adapt through migration may be five or ten times slower than we once thought. The conclusion of three different research analyses: tree migration rates are below those required to keep up with projected climate change. That leads to the scenario we may already be witnessing in the West: a catastrophic die-off with a delay of unknown duration before warm-adapted trees will move in, if they can at all.

THREE POSSIBILITIES could at least moderate the menace of a "big die-off" here. The most obvious is that in Virginia, as elsewhere, some trees are already at or near their northern limit, such as shortleaf pine, longleaf pine, American holly, and water tupelo. Given a limited amount of warming and sufficient rainfall, they could be reasonably happy with their new situation and could even expand their ranges in landscapes that are not too fragmented. If the climate continues to grow hotter, even their tolerances will be exceeded.

Then, too, there is evidence in the pollen record of at least a few places that were refuges, where plants and animals were protected enough from the harsher aspects of the ice age that they could "hide out" during the centuries that passed, until the climate became more hospitable. In the rough topography of mountain Virginia during a hotter epoch, "you actually have an advantage, because species can shift range upslope," the conservation biologist Reed Noss told me.

They may also shift to a slightly cooler "micro-environment," such as a slope that faces away from the sun, or a place closer to a spring or a seep. These small refuges seem to have preserved a variety of species in the mountains during past climate changes, so they might shelter core populations of plants and animals until the climate cools again in some far-off era. In the Shenandoah Valley, for example, sinkhole ponds in limestone formations have preserved both northern and coastal plain species over a ten-thousand-year span. Similarly, shale barrens in the mountains serve as refuges for western prairie-like species.

We can expect extinctions because of the advancing heat just the same, Noss added. Birds are mobile despite some of the barriers we put up, but bears are impeded. Salamanders need constant moisture. Virginia spring wildflowers such as trillium or toothwort, for example, are vulnerable because they are slow movers — they rely on ants to move their seeds to new locations.

A final possibility for avoiding a long "dead zone" transition period until the forests recover arises from what Jacquelyn Gill, a paleoecologist and biogeographer at Brown University, calls the leading edge of research on tree migrations. It is a debate that aims to resolve a problem called "Reid's paradox." When scientists look at how trees reproduce — the mechanics of seeds and wind and dirt and pollen — the rate is far too slow to explain how quickly they seemed able to go north as the ice age withdrew. These tree species just do not normally show an ability to migrate nearly as fast as the pollen record seems to indicate that they did.

The deus ex machina proposed by some studies is animals. Squirrels, blue jays, passenger pigeons, or mastodons may have transported the seeds over longer distances, explaining the rapid dispersal that the pollen record shows. Other research argues for those refuges — small groups of trees that persisted in safe areas surprisingly near the edge of the ice sheet, then expanded when the climate warmed. The evident presence of these trees may have fooled researchers into thinking that they migrated in, from a long way south, and quickly.

Gill said that neither explanation bodes particularly well for the forests in a fast-heat era, however. Passenger pigeons and several other seed-transporters are extinct, and the numbers of most surviving species are severely diminished. Refuge spots are few, since so much of the landscape has been converted for our use. "I would say, yeah, the paleo record as a whole is telling us that — particularly in this world where large animals are few and the habitat is very fragmented — it is not a good scenario to be a tree and try to get to where you need to be in the next five hundred years," she said.

For that reason, conservation scientists have been increasingly embroiled in another debate, over what is called assisted migration. Unfragmented landscapes through which migrations can occur are almost as rare as the mastodons, but there is one animal that could act as an intermediary for moving trees and other plants: us. We might physically relocate as many species as possible to help ensure their ultimate survival. If that seems harebrained or overambitious, supporters of the idea say, consider the alternatives.

Shugart measures the problem efficiently: "Let's say the climate changed tomorrow afternoon. Your smart move would be to head off down to Georgia someplace and get a whole bunch of plants that would grow in our forest in this new climate and hire every high school kid on the planet to plant them.

"But it's still going to take a couple of hundred years to develop the new forest. Even in ideal circumstances, there's a delay. You can only push the succession process so fast, which means you're going to end up for human-lifespan time periods with plants that are either going to be dying, or at least not prospering. It's still going to be a couple of hundred years before that stuff's in place. You'd have to do something to the existing forest to get new species to repopulate, too. How are you going to plant Georgia pine trees in an oak forest? Well, you have to knock the oak trees down, to start off."

The possibility of large-scale intervention of this kind in order to adapt the forests to global warming has provoked heat of its own. Mark Anderson, director of conservation science at The Nature Conservancy, said: "There's no agreement on it at the moment. Those of us like me who are pessimistic about climate change think we've got to let go of the purism and work with nature" by relocating some plants. "There's a large voice for that at this organization," he said. "There's another group that is still extremely uncomfortable with it. Because the track record of people intentionally monkeying around with introductions and movements is really poor. And we tend to have created more problems than we ever created solutions."

SCIENTISTS TRYING TO figure out how to preserve as much of the natural landscape as possible have stressed the need to save the "connectivity" that still remains along a northward track. This would allow for the possibility — the hope, really — that plant and animal populations can migrate quickly enough to stay within their heat tolerances.

These corridors figure in another early-stage plan. The Virginia Department of Conservation and Recreation's Natural Heritage Program manages 55,000 acres and 400-odd rare species and natural communities in a network of preserves. Its staff is slowly mapping potential migration pathways, especially for the most vulnerable species.

The James River spiny mussel, for example, already one of the rarest animals on the planet, will have trouble as the river warms. A high-elevation salamander that exists only on Shenandoah Mountain has no more "up" to move to. Sensitive joint vetch, a freshwater marsh plant that appears on all the endangered lists, is now rated "extremely vulnerable" to climate change. Those freshwater tidal marshes are themselves threatened by sea level rise as it pushes salinity up the rivers.

The reality, as program director Tom Smith describes it, is that while climate change is at the top of every conservation biologist's agenda, "our climate initiatives are pretty paltry. The list is short. We're dealing with those threats to biological diversity that are hitting us every day," he said, such as invasive species, and the conversion of natural land cover to agriculture, tree farms, houses, stores, and asphalt.

"Times are tough in state and federal government, and the resources are not there to do serious planning for what's coming in ten, twenty, or fifty years," Smith said. There's not even enough money to sample the possible impact of climate change, right now, at the program's network of four thousand monitoring sites around the state. They are observation stations awaiting observers.

"We've always been in the losing game business," Smith adds with a laugh. "Climate change is a really bad thing confronting us, but we've always been looking at an enormous challenge. There's nothing else we would rather do, so we've spent our careers being hopeful and figuring out the best path forward.

"We certainly have some preserves where rare species may be lost to climate change. The anxiety is that though the places may always be interesting or even diverse, those species and communities aren't going to be there anymore. There is no active planning to make sure they won't go away. We really need more capacity to do that work."

Mapping and species evaluation projects have at least begun, though. Sometimes "it's a triage kind of deal," Smith added. "In a way, we're looking for organisms and communities that aren't going to make it, so we don't spend money on them."

Virginia does have an exemplary tax incentive program that promotes conservation easements among private property owners. It has protected tens of thousands of acres of landscape. If those tax deals could be yoked to the pressing need for wildlife corridors, it would offer a chance, at least, to ease some of the effects of climate disruption.

Smith also finds inspiration in a small-scale project pursued by his agency and the state Department of Forestry. It explores the possibility of expanding the range of longleaf pine, a warm-adapted species that once covered a million acres of the state. Today, only 180 cone-producing trees are left here. The work is promising—an early blueprint for forest relocations that may sometime soon be called for. "We've planted about 180,000 Virginia longleaf seedlings on natural area preserves in southeast Virginia. So we're managing for a habitat loaded with rare species that should continue to do well in the face of climate change," he said.

Nature Conservancy scientists have also wrestled with how to plan for climate change on their portfolio of preserves, which includes some 300,000 acres in Virginia. They were acquired by the non-profit to protect rare species and vulnerable ecosystems, on the assumption that the climate would be stable. As the climate alters, some of those preserves will become mausoleums instead.

The Conservancy's most recent work is to recast the strategy of connectivity. It isn't enough just to make sure migration corridors are not blocked by cornfields, superhighways, or strip malls. It won't work to hope that wetland species, for example, can somehow migrate across shale barrens or talus slopes, or to assume that coastal dwellers can somehow migrate up and over high mountain ridges. The need is to connect landscapes that offer the right combination of soils, elevations, and climate, so that organisms adapted to those conditions have a better chance at survival. This is called "ecological flow."

"We literally cannot predict at this time where everything's going to go, and which species are going to make it where," Anderson said. "What we can do is focus on conserving those different environmental stages that will support these actors, these different sets of species. I think we're confident that we've identified the best places to focus on, but we're not entirely confident that those places will stay resilient. It really depends on what the climate does and what the people do."

The Conservancy has mapped lands that are still available for conservation, still exhibit some "flow" potential and some resilience — the capacity to adapt to climate disruption. For Virginia, here's how that looks. In figure 15, blue shows areas of concentrated ecological flow, and green shows areas of ecological resilience. These are the lands that, if we plan intelligently, we will make certain are not further built upon, fenced, roaded, mined, logged, farmed, or otherwise altered as climate disruption advances. The fragmented look of these maps is a stark challenge, but the opportunity they represent is striking, too.

Reed Noss, the conservation biologist quoted earlier, proposes a crash conservation program for maintaining migration corridors in the Southern Appalachians. "In general I don't like to go on record as a proponent for condemnation of land as a blanket solution, but I think it has to be part of the solution. We've done it for dams and reservoirs and highways, why not for natural areas? The government has done virtually nothing on land acquisition. Everything has its price, and we could just buy land," he said. Noss has calculated that for what has been spent on the Iraq War alone, a million square miles of land could be purchased and safeguarded for its environmental value to humans and to wildlife.

On the millions of acres of national forests in Virginia and the rest of

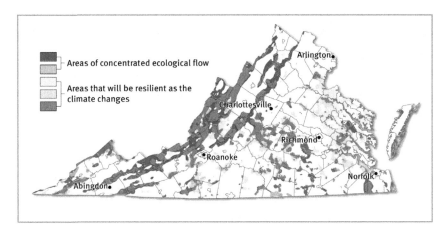

FIG. 15. "Ecological flow": paths where plants and animals could migrate as the climate heats up

the region, Noss recommends an immediate halt to road building, and ripping up and revegetating existing roads wherever possible. "We should stop using heavy machinery that disturbs the soil, and undertake only very low-intensity timber harvests," he adds, in order to preserve the closed forest canopy as a buffer against the effects of warming.

The climate change scientist Dominique Bachelet says that water resources may be unreliable in the mountains as the climate warms. If we want our forests to be better adapted to future drought conditions, she says, we have to reduce the number of individual trees and the number of invasive species. They suck up water and add to fire danger. "Leave the big trees to provide a shady canopy," Bachelet advises — don't log them anymore, but thin out the "dog hair" forest of skinny young overcrowded trees.

In New Jersey's Pinelands, warming winter temperatures have generated a population explosion among southern pine beetles, and tens of thousands of acres of trees have already been killed. The infestation is moving northward. Happily, in Virginia's forests there is no readily discernible evidence so far of the impact of the warming climate. Southern pine beetles, among the most destructive insects in this region in the past, are in abeyance for now.

Tree species monitored by the U.S. Forest Service and others show no signs of migrating north, beyond their historic limits. There is also little evidence that tree species are migrating in the areas of more rapid warming — the Northeast and Upper Midwest. But that is a mixed message. Species that fail to adapt to the warming climate by moving are even more at risk.

And the relatively placid status quo for Virginia is not inconsistent with what has happened in the West — healthy-looking forests that suddenly succumbed — or with the modeling of Virginia's future forests under projected climate change. "Which is tough on people who are trying to make decisions about land," Shugart said. "Because it means that everything's going to look just peachy, and then all of a sudden it starts changing. And by the time you've noticed it changing, it's over. It's not giving you a nice clean signal that 'Gee, I'm not feeling too good here.' It's: poof!"

Among the hopeful suggestions that arise in global warming discussions is that CO_2 may help the growth of forests. There is evidence that in the Far North, at the edge of the Arctic tundra, it does. At Duke University, the lab of the forest ecologist James Clark mounted some of the earliest and largest experiments, nearly two decades ago, to test the possibility that more CO_2 in the atmosphere could lead to more tree growth here in the Southeast.

"There's no question that CO_2 concentration of the atmosphere is really critical for plants," he told me. "They're limited by the amount that's in the current atmosphere. If you add more, they grow faster. If you reduce it, they struggle." So his Duke team fumigated forest stands with CO_2 over varying periods, but the experiments showed that, in general, any enhancement of growth in our latitude is limited. "It's not going to cause a big stimulation in forest production, that's pretty clear," Clark said.

Another potential positive outcome for warming and forests would be that the growing season is longer. Southeastern forests produce more wood than many other regions for that reason. But because the climate has been comparatively stable for thousands of years, our forests are well adapted to take advantage of the climate we have, Clark said. The research so far has not, unfortunately, found that trees are able to adapt to a new climate and prosper when growing seasons are artificially lengthened in a lab.

Clark and his colleagues have been following the life histories of nearly fifty thousand trees in the region, some for as long as twenty years. Their responses to stress differ widely. For example, the common hardwood sweet gum is resistant to summer drought. Loblolly pine, a key species for the lumber industry in the Southeast, is quite vulnerable.

"The character of southeastern forests will really change," Clark said. "We're going to be facing a very different climate. I wouldn't put moisture-demanding plants in your garden. I'd think about things that can really deal with the droughts.

"The precipitation we can't really predict, so maybe it'll be extremely wet. But in most scenarios, you have a situation where even though the rainfall goes up, moisture is still a big issue"— erratic, all-at-once rain, with possible long periods of heat and drought during summer and fall. The "big die-off," Clark acknowledged, is not an unlikely outcome.

The expanses of Virginia forest that we've had the luxury of taking mostly for granted need thorough reconsideration. Looking back over the transformations wrought since Europeans arrived here, Shugart was optimistic about what people can do if they turn to it: "I mean, we sure as heck have seen incredible land-use changes at the hand of man — perhaps in directions I don't particularly like. But we were pretty good at banging down a forest with our hands when any logical person looking at it would have gone back to England. You know what I'm saying?"

THE FIDDLERS

I've found the place to "build the home of your dreams!" It's described just that way on the real estate listing here in front of me, and on countless others for Virginia waterside properties. They are a dreamscape, indeed: a queasy mix of sleep, hope, mirage, paralysis — life on a momentarily serene yet precarious view lot.

That prospective half acre is on Sandfiddler Road in the community of Sandbridge, near the southern limits of Virginia Beach. It's for sale just now at $699,995, and it faces the Atlantic, right on the beach with an ocean view, from a few feet above sea level.

There's a serpent in this Eden, though. Well, actually in the front yard of the whole neighborhood — a twisting, roiling green monster, on occasion. The sea, which erodes the beach nonstop here, has to be fended off every so often with a "replenishment"— 2 million cubic yards of sand dredged up from a long way offshore.

The wider significance of that battle, and what makes Sandbridge's troubles emblematic rather than isolated: climate change is rerigging the geomechanics that govern the level of the sea, and where we can and cannot live along the 10,000 miles of Virginia's tidal shorelines — the longest of the lower forty-eight states. Much of it will resemble Sandbridge, and may be engulfed in the future.

Though it sounds fixed, there is always a shifting arithmetic behind the idea of "sea level" because, of course, it isn't. We can start to think about this by using what is sometimes called the bathtub model: the water in the tub goes up or down, and maintains the same elevation relative to "dry land" all the way around. So if we add more water to the oceans, thinking in terms of this model, the effect is the same on all the world's shorelines.

Instead, though, sea level depends on several factors, starting with the shape and elevation of the land. Because so many of us live so close to it and have adapted our homes, livelihoods, commercial buildings, transportation, and utilities to it, a confident sense of where the water stops is crucial.

Seventy-eight percent of Virginians live within 20 miles of the Chesapeake Bay, or the Atlantic, or of tidal rivers. But sea level, and the history we've used as a guide, are no longer stable. Already, flooding is a constant expectation in some places, especially on the Eastern Shore of the Delmarva Peninsula and in the Hampton Roads area.

To see what's already occurring and what's predicted for the coming century, let's complicate the bathtub model a little. If you run the water until it's about half-full and then climb in, you have plenty of room to slosh around before your floor, and your chin, get wet. But you have to be careful just the same not to make many waves, or at least not big ones.

Imagine, though, that your tub is the ocean, the tidal rivers, or the Chesapeake. Twice a day, high tides come in, adding to the water level and reducing your margin of safety. Along Virginia shorelines, the tidal range — the difference between high and low tide — can vary from a few inches to more than four feet.

Twice a month, the full moon and new moon show up, and, with the sun, they pull the tides significantly higher — maybe 6 or 7 inches. That's okay. We build with these "spring tides" (which have nothing to do with spring) and normal waves in mind. But atop those regular, predictable variations in sea level comes a list of sporadic ones such as storms, and they carry in even more layers of water.

During the tropical storms that can be expected annually, on average, we add about 3 feet to sea level, as their high winds push larger waves onshore. A so-called "hundred-year storm" — each year, there's a 1 percent chance for one of those to occur — can add an additional 4 feet on top of the rest.

The biggest storms, especially hurricanes, produce not only huge waves but also storm surges, and how much damage they cause depends on their angle of approach, among other factors. Surges are huge mounds of water that build up under the pressure of the wind and move landward.

Big storm surges have been measured at more than 9 feet at the Virginia shoreline. Unless we've taken precautions, the tub's going to spill over in a big and damaging way at this point, and we may be in up to our nostrils, or higher. A hurricane arrives somewhere in Virginia every 2.3 years on average.

Yes, you may think, but that's just scaremonger rhetoric — it's imagining the worst case on steroids: a full-moon, high-tide, broadside hurricane. When would that ever happen? As I wrote that sentence, though, a rain-heavy hurricane named Sandy was tracking north from Florida. We had a full moon through several high-tide cycles as it approached and worked its way along the coast.

This bad trifecta isn't mere coincidence. Virginia was mostly though not completely spared what New York and New Jersey had to suffer on this occasion, but our odds of enjoying that kind of luck for long are, alas, just not good.

In fact, they shorten each year as the sea moves forward. Global warming is steadily making the language we have used to describe our climate obsolete. As it redefines sea levels, it also changes hurricane threat levels, and what "flood-prone" means. Sandy was designated a storm, not even a hurricane, when it made landfall and set records for damage.

Hurricanes have havocked the mid-Atlantic coastline and its inhabitants since the first Paleo-Indians set up camp here, but the concern now is that climate disruption, even aside from sea level rise, will make the impacts worse.

Research on whether it may bring more frequent hurricanes is still inconclusive, with contradictory factors at play. But several studies indicate that whatever happens with hurricane frequency, their intensity can be expected to increase as the oceans heat up, adding energy to the climate system. Assuming the same total number, there will be more chances for Category 4 and 5 hurricanes in the future, and lower odds for the weaker ones, than at present.

STORMS AND SEA LEVEL rise offer one more reminder that it's useful to talk about climate change on a regional scale, in terms of adapting to its impacts as best we may. But the warming itself, the ultimate causal factor, is planetwide. Virginia shorelines are more than 2,000 miles distant from

Ilulissat Harbor, Greenland, for example, but their fates are linked. There, 150 miles north of the Arctic Circle, tourists arrive in increasing numbers to witness the warming planet as icebergs from the melting, crumbling Ilulissat Glacier — Greenland's largest — float down the fjord.

Each year, the front of that glacier is retreating nearly 2,000 feet, and it loses about 50 feet of its thickness. Recently, its rate of flow toward the fjord has more than doubled. This river of ice drains the mile-deep slab of the Greenland ice sheet. It is the last remnant of the ice-age mantle that dominated Virginia's climate twelve thousand years ago.

The glaciologist Waleed Abdalati of the University of Colorado's Earth Science and Observation Center has spent several years doing research in Greenland and on Ilulissat. His teams use monitors and measurers bolted to aircraft, snowmobiles, and satellites to detect change along the immensity of the ice sheet. Sensors of its reflectivity and surface melt, thermal imaging systems to capture its temperatures, laser altimetry for height, ice-penetrating radar for thickness, interferometry to calculate velocity, gravity meters to sense the dwindling mass.

Among scientists, Abdalati told me, the net result of all that labor is "a consensus that we expect sea level to rise, and we expect it to rise at an accelerating rate. For a lot of us, 3 feet of sea level rise certainly seems plausible by the end of the century. But there are some who look at the last time the earth was as warm as it is now. Then, oceans were 15 to 18 feet higher than today. So there's some concern that the sea could rise quite a bit more."

Those melting ice sheets in the Arctic are adding a roughly estimated 50 cubic miles of water to the oceans each year and altering their salinity. And as global warming heats the ocean water, it expands. Both factors contribute to sea level rise in roughly equal measure, for now.

"The Greenland ice sheet is shrinking more rapidly than we had thought it would in the past," Abdalati said. "We used to think ice sheets responded very slowly to changes in climate. They're big. They have high inertia. It takes a long time for that heat on the warm surface to penetrate into the ice and make enough of a difference to change its characteristics. But what we found is the ice sheets actually respond very rapidly to climate change in a structural sense." For example, as Greenland ice gets warmer and wetter, it is more slippery and spills toward the ocean more quickly.

As his colleagues refine their models of glacial melting with increasing urgency, "the only rumblings I'm getting are that we're not expecting some kind of horrible runaway effect, which would make for the worst numbers. But we still don't really understand the ice, either," he said.

The recently retired NASA climate scientist James Hansen warns of the possibility of much faster melting and sea level rise, and has called for a crash international program to firm up projections. He has also criticized his colleagues for being too reluctant to alert the public to their research findings about warming and the risk of impending jolts — such as the meltdown of polar ice — out of fear of compromising their credibility in a highly politicized arena.

Abdalati, who has also served as chief scientist for NASA, said he is ambivalent about whether he holds back much. Sometimes after he gives a speech, "they tell me I'm not scary enough. But other times they say I'm too scary," he told me.

"You know, a scientist has to be careful as to whether we are informing or persuading. And my experience has been that when we get very extreme, we tend to push away the people that are never going to buy it anyways, and keep in our camp the people that are already there, and it all makes talking to the remaining middle group a little more difficult. I have found that by being very open with what we don't know, by explaining what the uncertainty is and then sharing what I think and why, people are willing to enter into a conversation with me."

Odd to think that the meltwater he observes cascading from Ilulissat, half a world away, registers here in Virginia's part of our shared tub. It shows up, though, on tide gauges at places like Kiptopeke Beach. In both places, the gearwheels that drive sea level rise — greenhouse gas emissions, a warming planet, melting ice sheets, and the expansion of ocean water as it heats (same as any kind of water) — are engaged, and accelerating.

Maybe you'd expect a tide gauge to be a seaweed-slick yardstick, nailed to a piling at the end of a pier. There is some of that kind of measurement in NOAA's system. At the end of a pier south of Sandbridge, for example, tidal benchmarks are notched into concrete posts, to mark sea level during various periods over the last few decades.

But Kiptopeke State Park, on the southern tip of the Delmarva Peninsula, has the updated version. Mounted on the far railing of a pier that often fronts a horizon of blown-out whitecaps, its plastic pipe extends down into the water. The pipe is fitted with an emitter that sends high-frequency sound waves down toward the undulating surface. When they hit, they bounce back up the pipe to a sensor, and it calculates the time that that short round-trip takes, which in turn tells how high the water is. A continuous reading, once each second, is taken as the water level responds to wind, tide, current, the ripples of passing motorboats, and, over time, the warming climate.

The information is averaged by the electronics in a squat white shed nearby, then forwarded up an antenna and on out through a satellite link to a NOAA data-processing lab. Kiptopeke's station, part of a national coastal network, began reporting in 1952. Another station, across the bay at Sewell's Point, Norfolk, has recorded tidal history since 1928, and the one at Glouces-ter Point, near Yorktown, since 1951.

They all tell us that sea level has been rising in Virginia about 1 inch every seven or eight years—a foot in the last century. The rate from the Chesa-peake northward is sharply higher than for points south, according to John Boon, a longtime sea level rise researcher who is also a professor emeritus at the Virginia Institute of Marine Science (VIMS).

Some part of that is because coastal land, especially around Virginia, is sinking over time, bringing the sea farther inland as it does. When the con-tinental glacier expanded south as far as Pennsylvania during the last ice-age freeze-up, the land ahead of it, including what is now eastern Virginia, was forced up by the edge and weight of the ice, as if the heel of a giant hand were pushing on a blanket. It created what geologists call a "forebulge" in the Earth's crust. The epoch ended, the ice retreated, and the bulge has been sinking.

Decades of pumping out groundwater for industrial and other uses has made the land sink even more in some localities. The continuing after-effects of a meteor crater from a long-ago strike near the south end of the Bay have also been implicated in the subsidence.

And Boon said that one more factor may add to sea level rise. There is evidence that the patterns of Atlantic Ocean currents pile up water onto this part of the coast. "The acceleration I found, and that other researchers have found, seems to be related to changes in circulation in the North Atlantic,"

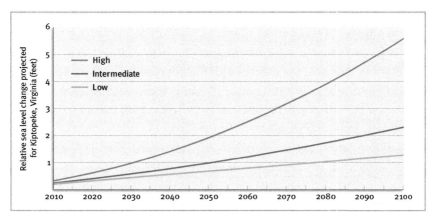

FIG. 16. How fast is sea level projected to rise along Virginia's shorelines?

he said. "This is a different process than melting ice or thermal expansion, although it's associated with global warming, too."

The U.S. Army Corps of Engineers is often called upon to build, or offer advice on, public works such as seawalls, roads, bridges, and bulkheads. The Corps has combined studies from the National Research Council and its own data to generate three projections for each tide station about where future sea levels may be. Figure 16 shows what they look like for Kiptopeke, and its curves are very similar to the rest of the Virginia shoreline. The lowest line is a straight extension of the current rate of sea level rise. You might think of it as the most optimistic case, but it isn't. Instead, that line is entirely mooted by a fact that the intermediate line acknowledges: analyses of historical tide data show that the rate of sea level rise in Virginia is accelerating. The line that reality describes is curved, not straight.

You can see that the intermediate curve forecasts 1 foot of sea level rise by the year 2050, and 2.3 feet by 2100 (compared with 1992). Just 1 foot of sea level rise would bring momentous and expensive change to tidewater Virginia. It would push salt water onto roughly 40 square miles of dry land, for a start.

The third, uppermost curve is the Corps' worst-case projection. John Boon spent the better part of a year analyzing tide gauge data from Key West to Boston to see which of the Corps' trend lines seems most plausible.

He was reticent about discussing sea level rise while the work was under way, and especially concerned about overstating the problem.

"People feel some of these warnings are perhaps a little extreme," he told me. "I think we have to be careful with that kind of thing and not be alarmist, but be certain of our estimates." When it was completed, however, his analysis showed that in Virginia, sea level rise has been tracking the Corps' worst-case projection closely.

As well-regarded as Boon's analysis is — it has been published in the peer-reviewed *Journal of Coastal Research* — it is still just a single entry in the research record. But two other independent research groups — Tal Ezer, of Old Dominion University, led one, and the late Asbury Sallenger of the U.S. Geological Survey, the other — have also produced peer-reviewed, published studies that corroborate Boon's result: accelerating sea level rise.

So the Army Corps' top curve is now much more strongly indicated as the predictor of where we are headed with sea level rise: it shows 2 more feet of saltwater at spring high tide by 2050, and 5.6 feet by 2100. That would have the same impact as a strong Category 1 hurricane storm surge, twice each day. It would push the levels of what today are Category 1, 2, or 3 hurricanes one level higher.

Some scientists would call that estimate unduly cautious, however. They would add another, more quickly accelerating curve of sea level rise to the chart, above the others. The book *The Rising Sea,* for example, counsels that the United States should be planning now for 7 feet of sea rise by the end of the century. The authors, Orrin Pilkey and Rob Young, coastal geologists at Duke and Western Carolina universities, call theirs "a cautious and conservative approach." A recent report on sea level rise and flooding for the Virginia legislature by the Virginia Institute of Marine Science cites 7.5 feet of sea level rise by around 2100 as its worst-case estimate, based on a federal technical assessment.

PROJECTIONS FOR climate change usually only take us to the end of the century. Beyond that period the uncertainties multiply, and anyway, we'll be a few generations down the line by then. Global warming and sea level rise don't end in 2100, however. They will continue for an indefinite period that

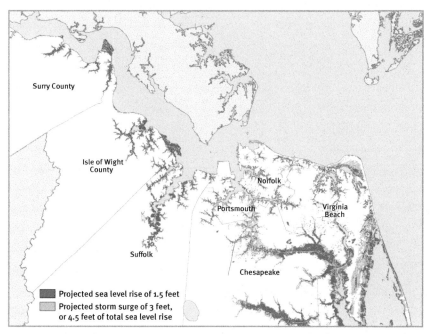

FIG. 17. Sea level rise and the Southside

may span tens to hundreds of years, even if greenhouse gas concentrations are stabilized.

In the much nearer term, the coming decades, any increase in sea levels will also push salt water upriver and into aquifers, destroying an unknown amount of freshwater capacity for drinking water and for industrial uses. (The U.S. Geological Survey has yet to study the potential extent of that problem in Virginia.)

The Hampton Roads area — Virginia Beach, Chesapeake, Hampton, York, Newport News, Norfolk, Poquoson, James City, Suffolk, Portsmouth, Surry, Isle of Wight — is most at risk, for two reasons: its concentration of population, the largest on the coast between Miami and New York, and its low elevations. Thousands more live on vulnerable shorelines on the Northern Neck, the Eastern Shore, and inland along tidal rivers.

Figures 17 through 21 were part of a recent VIMS study. They depict the impact of sea level rise, assuming no additional seawalls or other protection.

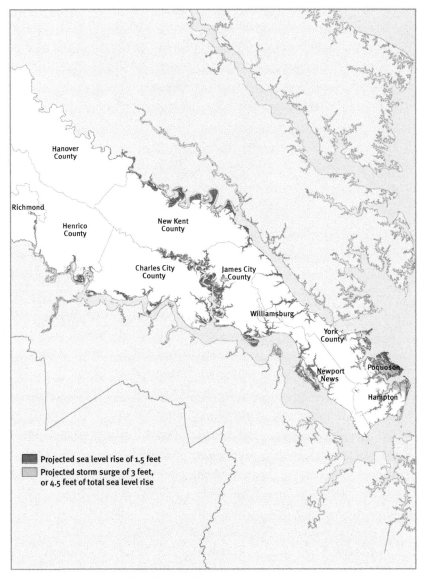

FIG. 18. Sea level rise and the Peninsula

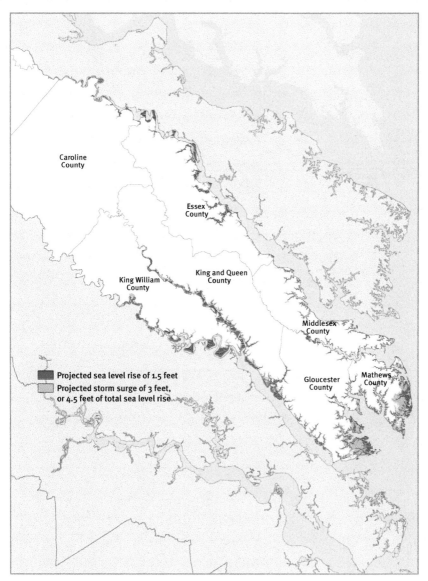

FIG. 19. Sea level rise and the Middle Peninsula

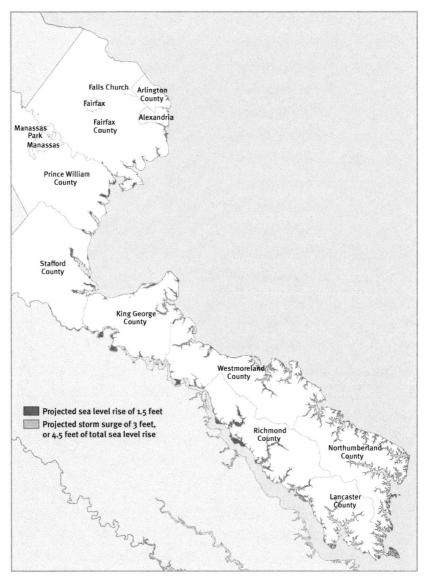

Falls Church
Fairfax
Arlington
County
Alexandria
Fairfax
County
Manassas
Park
Manassas
Prince William
County
Stafford
County
King George
County
Westmoreland
County
Richmond
County
Northumberland
County
Lancaster
County

■ Projected sea level rise of 1.5 feet
■ Projected storm surge of 3 feet,
or 4.5 feet of total sea level rise

FIG. 20. Sea level rise and the Northern Neck

FIG. 21. Sea level rise and the Eastern Shore

They show potential permanent inundation with 1.5 feet of sea level rise (due here before midcentury, according to the projections). They also show potential flooding with a 3-foot storm surge, in addition to 1.5 feet of sea level rise. Finally, the total of the two numbers indicates how much land would be inundated permanently with 4.5 feet of sea level rise (due here around 2090, according to the projections).

Here's a different kind of summary, making use of other research that has tried to approximate the economic, ecological, and infrastructure impacts of varying degrees of sea level rise.

TWO FEET

Estimates vary, but assuming just the current level of flood protection, 2 feet of sea level rise could inundate 82 square miles of dry land and 660 square miles of wetlands in Virginia. Fifteen miles of Virginia interstate highway would be submerged or at risk in storm surges, along with 46 miles of principal arterial streets, 54 miles of the national highway system, 52 miles of railroad, and 35 percent of the acreage of our ports and their shipping facilities.

Because transportation systems are networks, just one flooded section of road or rail can have far-reaching effects, as one federal infrastructure report notes. The nationwide logjams following a storm that closes a single major airport for a day or two are a telling example. The difference is that when roads, railroads, and port facilities are flooded or destroyed, they take far more time to rebuild or reroute away from the new coastline.

THREE FEET

The Hampton Roads Planning District Commission estimated what is at risk of being permanently drowned on the one hand or frequently flooded on the other. With 3 feet of sea level rise — projected to be here by about the year 2065 by the Corps of Engineers — the middling estimates are that 238 square miles of Hampton Roads' land would be inundated, and more than 112,000 people could be displaced. Their 45,000 homes, part of anywhere from $9 billion to $25 billion in property improvements, are at risk of inundation unless they're protected somehow. Add to the list 506 miles of roads and 2,000 businesses with about 25,000 employees.

These tallies only consider sea level rise at spring high tide, where its impact will be more or less constant. The real vulnerability is measured by considering storm surges and waves. They could add several more feet of trouble onto the ledgers. As the waters rise, the losses expand, safety zones narrow, and more land is vulnerable.

SIX+ FEET

As many as 628,000 Virginians live on land that is now within 6.5 feet of sea level. The population number is a rough estimate, but it includes about 572,000 people living in Hampton Roads, 30,000 along the Potomac, 16,000 on the Northern Neck, and 10,000 on the Eastern Shore.

So the numbers rain down, numbingly, as bloodless as confetti — the stuff of leaden reports, easily shelved. A brief notation from a firm that advises the insurance industry may bring us at least a little closer to the thing itself:

> Storm surge typically moves at 10–15 mph, keeping pace with the forward speed of the hurricane. The hydraulic impact created by these waves tends to be incredibly destructive because one cubic yard of sea water weighs approximately 1,728 pounds — almost a ton.
>
> Compounding the destructive power of the rushing water is the large amount of floating debris that typically accompanies a surge. Trees, autos, boats, pieces of buildings and other debris can be carried by the storm surge and act as battering rams that can damage or even collapse buildings in the waves' path.

The essayist Leon Wieseltier, who endured Hurricane Sandy, yanks us in closer still:

> This time the other people to whom terrible things typically happen included us. . . . [T]he waters of the bay to the north and the sea to the south rose angrily and, like armies of pillaging invaders, met to destroy. The flooding was vicious. The waters climbed six feet, eight feet, ten feet. They smashed in doors and knocked down walls. Brick and

cement crumbled before them. As the waters began to attack our mother's house, my sister phoned me, and I do not believe I will ever shake the sound of her terror. Our impossibly frail mother was upstairs, sweetly oblivious to the danger. The SUV was underwater and carried off into the garden. The phones were dead, except for the antiquated land line. The power went out. The live wires brought low by the wind started sparking, which threatened to ignite the fuel in the floodwater from the bay's broken boats. As the waters began to recede — they came to kill and then fled — the extent of the devastation was revealed. The nights were cold and black; we lived in dread of sunsets. The police created a checkpoint at the entrance to the neighborhood because looting was feared. We kept our mother warm with propane heaters and then with a gas generator, which also restored her hospital bed, though it required heroic dead-of-night searches for canisters of gas. We felt ruined, helpless, indignant, severed, confused. Our spirits were smashed.

POKER

Those not-too-distant horizons—and the difficult choices they pose—seemed nearer as I headed down Princess Anne Road one afternoon with the planner Clay Bernick, the administrator of Virginia Beach's Environmental Sustainability Office. Its 440,000 inhabitants make this city on the Atlantic at the mouth of the Chesapeake the largest in the state. We rolled east and south down a gentle gradient toward the water and talked about the sea level rise that is coming the other way.

"Most of us want to live here on the coast," Bernick said, "but in southern Virginia Beach your average elevations might run anywhere from 3 to 5, maybe 7 feet above sea level. Our coastal exposure is something like 30 miles of ocean shoreline. Then you include the Chesapeake and the other bays and inlets and tidal river shores. We have to worry about sea level rise just because of the geography of the city."

CoreLogic, an insurance industry consultancy, ranks Virginia Beach second only to New York among American cities in terms of potential damage to residences from storm surge. The city has also been ranked by the Organization for Economic Cooperation and Development, an international economic policy group, as one of the top twenty in the world in terms of overall assets at risk to sea level rise, vying with New Orleans, Calcutta, and Guangzhou. "Those are claims to fame that I think create some concern," Bernick summarized drily.

Virginia, he added, is somewhere near the beginning of planning for sea level rise: "Unlike our neighbors—both North Carolina and Maryland now have climate protection plans for each of their jurisdictions—Virginia does not have a plan." Studies have been undertaken, but Virginia's pace of real planning for sea level rise could be called "glacial"—except that most glaciers are melting so quickly the word no longer denotes slow change.

The Virginia climate change commission's report of a few years ago included a long list of recommendations for planning for sea level rise. The "action plan" included that the secretary of natural resources should take the lead in devising a "Sea Level Rise Adaptation Strategy" for natural resources, economy, and infrastructure. Years later, no such strategy has been drawn.

Fiscal conservatives take heed: sometimes delay is prudent. This time, it will prove enormously expensive. Virginia Beach, like any other city, must invest heavily in long-lasting public infrastructure: roads, water and sewer lines, bridges, public buildings, schools. A new bridge, for example, is typically built for a life span of eighty or ninety years. Sea level rise will make some of the existing infrastructure in low-lying areas—the kind Bernick and I were looking at through the windshield—obsolete, if it remains unprotected and unaltered.

This makes planning new infrastructure an increasingly expensive puzzle: How much will Virginia Beach be able to afford on its own, and who will pay the tab for the rest? Virginia Beach has no guidance from the State of Virginia about how to proceed in planning for sea level rise. Neither does the Virginia Department of Transportation offer guidance that incorporates sea level rise for its project engineers.

If we choose not to let the next Katrinas and Sandys carve out our future one disaster at a time and devil take the hindmost, there are several choices for coping more effectively. We can try to hold back the sea with many kinds of barriers. We can jack up structures above the water or try building up the land underneath them.

And we can plan a strategic retreat instead of a panicked one. We can abandon or allow only short-term building in some areas that will eventually be inundated, and protect only what can feasibly be saved. Each option carries consequences far into the future for private citizens, for our natural areas, and for public treasuries. Each requires long lead times, clear thinking, and extraordinary levels of funding to implement successfully.

If protection is a plausible engineering option—not always the case—then seawalls, beach sand, or other measures may be justified to protect high-value investments such as downtown Norfolk or the resort area of Virginia Beach. The cost of seawalls can range upward of $10,000 per square foot, though. We will be forced to relinquish land that is more sparsely settled or less economically vital, such as farms or residential neighborhoods.

There's an obvious political hazard in trying to abandon some parts of the landscape to the water, and paying to save others. Climate change is everyone's responsibility — and thus, no one's — so the responsible parties who should have to pay out the billions for protection are hard to define. The more immediate question now is whether to continue to incentivize even more commercial and residential construction and infrastructure in vulnerable areas. That is an investment of public money, and liability, in the eroding fantasy of living near shorelines that are no longer fixed.

"I think we're going to have to look at a variety of incentives, a lot of funding approaches, a lot of new land policies that may encourage retreat in some areas," Bernick said. In mostly agricultural southern Virginia Beach, a 3-foot rise in sea level may mean that 30 to 40 percent of the landscape disappears underwater or becomes wetlands or a hardwood swamp. "Will you be able to afford engineering solutions to protect that area, build levees and dikes and seawalls?" Bernick asks.

A few Virginia Beach and Norfolk residences have already been elevated out of harm's way, subsidized in part by the Federal Emergency Management Agency's taxpayer-supported flood insurance program. "Yeah, the stilts are coming," Bernick said. "One of the big concerns, though, is that you might be able to protect the structure, but how do you get to it?" The road, the water mains, the sewer lines will be underwater, and the electric lines will be strung on poles poking up out of the water.

In Virginia Beach's resort area, pouring dredged sand on the tourist beach and beefing up other shoreline protection measures may prove easier to finance through special tax districts or a room or restaurant tax. "I think we're going to have to dig real deep in the tool box and look for a lot of different options," Bernick said, "and I think we're kind of a microcosm in Virginia Beach of all those problems that other parts of Virginia are going to have to deal with."

Still miles inland, we were rolling through a landscape dense with newish buildings, 4 to 8 feet above sea level. Roadside ditches here are often full at high tide, and we passed skeletal stands of trees killed by saltwater intrusion. "Dramatic. Pretty dramatic stuff," he said. "I think people are going to be blown away by how much more vulnerable this will be."

We descended to a plain that was 5 feet above sea level. "All of this is relatively new development," Bernick said, looking out at multistory apartments and strip malls. Is there any way of preventing this, given the prospects? "No," he said. "The political will's not there to say no to the public." More apartments, only a few years old, came into view on the south side of the road. We saw marsh grass growing in the ditches.

"When it gets worse, the property owners will say, 'You let us build here, how are you going to protect us?' And we will lead the charge up to Washington, and try to get them to do it. There'll be residents before city council, and all the hotel owners and all the property owners down here will be out with pitchforks. To me it's not a liberal or conservative issue—it's strictly fiscal responsibility. Some places, like Norfolk, are even worse off than we are," he added.

IT'S A FAIR OBJECTION that portraying such densely urbanized places as underwater in the future is overblown, because we will come up with the staggering sums we'll need to build protection for at least some of them. But the reason to view Norfolk in these terms is not to pretend that it will all be engulfed.

It is instead to emphasize the point that the sooner we decide to act to save what we can, and withdraw from what we cannot, the more effective and inexpensive that process will be. It's a case of pay me now, or pay me later but with far more pain. Without comprehensive planning, we will be forced to fall back on irrational, reactionary approaches rather than well-planned adaptation.

As the examples above and in the next few pages suggest, delay will be wasteful almost beyond calculation. We'll spend heavily to see the new constructions and the ad hoc fixes, and the investment they represent, washed away. That squanders resources we will need for a more realistic and durable adaptation process.

Here's a different critique of the Norfolk-is-underwater scenario: it may be too mild a portrait of what's in store. The coastal geologists Pilkey and Young warn in *The Rising Sea* that "storm surge, storm waves, shoreline erosion, groundwater salinization, and infrastructure destruction will force a retreat from the shoreline long before actual inundation occurs. Simple maps showing the areas that will be slowly inundated by a given level of sea level rise should be viewed with great skepticism. They do not truly consider the inland region of the changes to be wrought by rising sea level."

I visited Norfolk with Skip Stiles, a local resident, the executive director of the environmental group Wetlands Watch, and a contributing author for the recent Virginia Institute of Marine Science (VIMS) study of sea level rise. He is frequently invited to participate on government planning panels, and served on the state climate change commission. He is a sympathetic tour guide for Norfolk's frequent-flood neighborhoods, with their salt-killed lawns and For Sale signs, as they enter a somber time.

"Sea level has been stable for thousands of years. We have nothing in our literature, our belief system, our legal system, our architecture or engineering related to sea level rise. You've got an ark myth, and you've got the Dutch boy with the dike, and that's about it. That's our entire cultural frame of reference for this. We're telling people sea level rise is coming, and it is going to eat your land and probably your city and people go, 'What? No! Norfolk's been here four hundred years. What are you talking about?'"

So far, because of that strange novelty and the sums required for engineering solutions—two studies recommended more than a billion dollars for dikes, pumps, and storm sewer repairs just for a start—Norfolk's approach has been piecemeal. The longtime mayor Paul Fraim had the courage to say publicly that the city may have to establish "retreat zones" at some indefinite time, abandoning sections that are exposed and where the fiscal burden has finally grown too heavy. Parts of the city are flooding nearly every month.

We stopped by a recently completed $1.25 million project intended to raise a waterside residential street in the Larchmont section 18 inches above the chronic floods. Several houses here have been jacked up on stilts, with FEMA financial help. The street still goes underwater at times, though. A mile away, plans have been drawn for a series of great steel sheets to be affixed to the supports under the Brambleton Avenue Bridge as a storm surge gate, to protect the homes and businesses behind it. It will be oppressively ugly, and the tens of millions needed to fund it have not yet materialized.

As the public expense of such fixes mounts, budgets thin. If localities pay, the conversation goes like this, in Stiles's rendition: "I don't mind helping somebody with a project here and there, but then, wow, this starts to be in my space, and my kids can't get a decent education. Then I've got a problem. I mean, these people whose property is going to be saved never invited me out to sit on their dock and drink margaritas."

If the state or federal government pays, the objections scale up to the national conversation: "Dude, what just happened, that you're taking all my money and putting it in a place where I don't even live?" The aching billions in damages from Hurricane Sandy have already put that question squarely in the middle of federal budget fights.

We made our way into a gracious old section of Norfolk, the most culturally evolved of urban landscapes to my eye, called Ghent. There's a Y-shaped inlet, The Hague, off the Elizabeth River, a light rail system, the Chrysler Museum of Art, and a long street of waterfront homes. A serene feeling pervades, but that's just a tourist's delusion. There is real distress here among property owners, and their losses are not somewhere safely far down the calendar.

A Unitarian church, more than a hundred years old, commands the corner of Yarmouth and Grace Streets. It resembles a fortress and probably is one for many of its 250 or so congregants who do not want to leave. Frequent flooding makes it increasingly hard to stay. On their website: "WARNING: Parking near the church during unusually high tides can severely damage your vehicle. . . . [there is] routine tidal flooding with brackish water, which is highly corrosive to vehicle parts."

The prospect of floods is unnerving, and when they come, they are expensive. There is worry that the city will stop issuing building permits in the flood zone, and that when the time comes to move, no one will buy the building. If there's an unusually high tide — not even a storm, just the wind or the moon — the church is cut off from the road, and that happens on average a couple of times a month, sometimes for several days. "Not only does the water come across the surface of the ground, but it comes up through

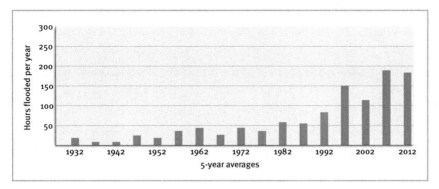

FIG. 22. Flooding has increased steadily in Norfolk's Hague section during the past eighty years.

the ground because this is all infill," the minister, a Richmond native named Cyndi Simpson, told me.

The congregation is attached to the building and its beautiful sanctuary, however. "The people who don't want to leave here have anguish over the thought," Simpson said. The congregation, not the minister, decides whether and when to vacate. "But I have an opinion: we need to get out of here, the sooner the better," she said.

BUYING A HOME is usually the largest purchase and the biggest risk most of us take during our lifetimes. There is nothing in Virginia real estate law, however, that requires warning a prospective home buyer about sea level rise. Nearby, a meticulously maintained, beautifully appointed home was for sale, at a figure nearing three-quarters of a million dollars. I inquired. After e-mail exchanges in which water was not mentioned, I finally asked about flooding. "During high tides when we have west winds, full moons, etc., water will come in the yard," I was told. "It has not been in the house. This is true of areas like Larchmont, Colonial Place, Ocean View. Parts of Va. Beach, Suffolk, Franklin. . . . Flood Insurance will be required." No mention that the water's coming closer each year, though it is possible the broker may not even have known.

"I know these people," Stiles said, looking down the block. They call him for forecasts of tides and winds when storms approach. "I know they're absolutely distraught," he said. Along this parade ground of the grand old homes of Norfolk, one good friend of his was inundated only four months after moving in. "She is simply beside herself with all the flooding," he said. "Is this the storm? Do I move everything to the second floor?" She knew she needed flood insurance. She was not told that flooding occurs, routinely.

Home buyers in many parts of tidewater Virginia could easily be victims and then later, potential practitioners of that old game of chance played with a marked deck, the Greater Fool Theory. "You know," the McLean attorney John Altmiller reminded me, "real estate doesn't always bring out the best in people, and it kind of can make them a little bit crazy. So you just can't count on people to do the right thing."

At some point, many will be left "under water," pinned beneath the weight of a worthless property. They'll hope or demand or litigate, on the basis that the government, which allowed such transactions, should step in with a solution. Here we enter into an unspoken agreement in which profits, especially from new construction, are privatized, but the losses are "socialized"— the public is asked to pay. The greater fool becomes us, because we did nothing to stop the game.

What is the responsibility, and the liability, of earlier sellers along the chain? Altmiller illuminates: "Here is my nutshell description about the duties of a person selling real property to another person in Virginia. Absent specific contract language, there is no duty to disclose anything about the property condition at all. The law in Virginia, as it has existed probably since people were getting off the boat in Jamestown, is caveat emptor, let the buyer beware.

"You're not allowed to throw a purchaser off his or her guard, and you're not allowed to lie to them. But if I'm aware the ocean is rising and bad things are going to happen, and I know this with a pretty strong certainty, do I have a duty to disclose that? No." A real estate broker has a duty to disclose material adverse facts they are aware of, he said. That creates a situation in which you might want to just fire your listing agent if he or she knows too much.

The law does require disclosures about lead paint and a short list of other defects. For notification about sea level rise, "There would probably have to be a conscious decision by the legislature to create a disclosure obligation where one presently does not exist," Altmiller said. "I imagine if you were trying to get something like this done as a matter of policy, of course you're going to deal with people who don't believe in climate change, as amazing as that sounds. Some of them are elected representatives."

Skip Stiles is worried that when the problem has become too big to hide, "panic disinvestment" will take over, resulting in chaos for wide swaths of Hampton Roads and beyond. Meanwhile, permits for construction are still issued in areas where a planned retreat would make far better policy, and several federal agencies still subsidize the insurance and the public works that enable the folly.

"I used to play poker a lot with my dad," Stiles said. "When somebody raises you, they say, 'the price of poker just went up.' The stakes are higher now. We know more about the risk, we know more about the cost, and the stakes have just gone up."

HAMPTON ROADS is usually the focus of discussion when sea level rise is the topic. It is the largest population center at the greatest risk from sea level rise outside of New Orleans, according to NOAA—yet another haunting competitive ranking of who is in the most ominous fix.

But the long shorelines of Virginia's tidal rivers, leading far inland, are also exposed to storm surges and rain-fed flooding. Surprisingly, the tide range often increases upstream along tributaries. The difference between high tide and low tide near the mouth of the James River at Newport News is about 1.8 feet, for example. Seventy-eight miles upstream, at the Richmond Deepwater Terminal, it is 3 feet.

Every hurricane brings the possibility of torrential inland rains, and a storm surge riding in on the back of the tides can meet a rain-swollen James, or Rappahannock, or Potomac, coming the other way. When the Potomac responds to sea level rise, it will push water landward into Arlington, Alexandria, and many other river towns.

Washington, D.C. (which, for discussion purposes, I will annex to Virginia), illustrates the present reluctance, even for public works of huge consequence, to incorporate sea level rise in planning. "The District's location at the confluence of the Potomac and Anacostia Rivers, combined with three buried waterways, broad floodplains, and relatively flat elevations, renders it highly susceptible to periodic flooding," a National Capital Planning Commission report states.

"Urban development has increased impervious surfaces, reduced vegetation coverage, and further exacerbated flooding and stormwater runoff through the entire watershed. This problem is especially acute in the National Mall area, given its downstream location." A few years ago, heavy storms led to the flooding of the National Archives, the offices of the IRS, and other federal buildings, some of which were closed for months.

A new $10 million pop-up floodwall has been constructed near the Mall. It's designed to be assembled quickly to block Seventeenth Street in the event that torrential rains cause the Potomac to spill into the city and threaten to swamp the many national landmarks along the Tidal Basin. At 10 feet, the wall is high enough that it should be able to stop even monster floods like the one in 1942 that caused what would now be tens of millions of dollars in damage. But the Army Corps of Engineers told me that the wall's design, in which it had a role, doesn't take sea level rise into account despite dozens of climate studies on sea level by near-at-hand federal agencies such as the EPA, NOAA, the Department of Transportation, and, oh yes, the Army Corps of Engineers. Just as the tides lift and lower the Potomac each day, so sea level rise will make a big difference when major storms hit. Every increment of sea level will reduce the D.C. wall's potential effectiveness against extreme flooding, making the most irreplaceable swath of the capital more vulnerable to the largest of storm surges.

James G. Titus, whose office is ten blocks from the Mall, has long been the EPA's presiding sea-level-rise Cassandra. He is the principal author of a long list of research reports and planning documents, and sea level rise has been his unwavering focus since 1982. "Essentially, I was told to create a program devoted to determining the possible impacts of the rising sea level and to find those instances where it makes a difference," he told me. "Even though, especially then, it seemed like something that was way off in the distant future, and wasn't at the time gaining a whole lot of interest."

The government studies of sea level rise's potential impacts over all those decades include cost estimates for seawalls, wetlands protection measures, dozens of coastal elevation maps, and "delphic Monte Carlo" statistical analyses of where the water may go.

Titus likes to windsurf along the Chesapeake, and his own family beach cottage is on the Jersey shore, about 150 feet from the water. He raised it 2 feet a few years ago, drilled drain holes in the floor of the storage shed, and bolted it to its corner blocks. That helped alleviate the wrath of Sandy. Only his daughter's toy wagon floated away.

He has made countless visits to city and county officials in Virginia and neighboring states over the years to explain what their communities should expect. "This was like trying to sell a supercomputer, not a PC. You are not expecting an immediate sale," he said.

Titus doesn't think that planning for sea level rise has to wait on still more studies of the problem. His rhetorical questions: "Do you need a cost-benefit analysis to tell you that you're going to protect Manhattan? That you're not going to allow the Jefferson Memorial to go underwater? That Miami is going to continue to exist?" And alternatively, it makes intuitive sense that farmland gradually being inundated now will probably continue to be. To try to protect it is too expensive.

Then there's the alternative of raising up a whole area of land. He once speculated, in a musing tone: "It's feasible to elevate a square mile of land. It's not that hard to estimate the cost of bringing a truckload of dirt to a place. You might have to go pretty far inland, when your sources are exhausted, maybe excavate a couple of mountains. And if that sounds radical, remember: how long have the natural elements been excavating mountains to elevate the coastal plain? So, we'd do it with dump trucks instead."

The remaining option, "strategic retreat," carries a clear short-term political liability: it would slow the building wave of coastal real estate "improvements" in Virginia and elsewhere. During his talks with shoreline communities, Titus said, "it was clear that homeowners and developers see little incentive to stop building on the coast, even if they did believe our forecasts. Why not build, if you're allowed to, and you can make a return on your investment before the water arrives?"

This sets up continued, mounting expense as the years pass. It is antiplanning rather than planning, a stubborn insistence on dealing ourselves a bad hand and worsening our losses. In newly built-up areas, according to a recent VIMS report, "affected citizens will pressure local government to 'fix' a problem for which there are few permanent solutions — all of them complex and involving considerable cost to the taxpayer. Smarter planning will see coastal managers and planners anticipating rather than simply reacting to these problems."

Titus traces the line of logic: "If you're a smoker, there's never any reason to quit today, because the probability of getting cancer will be about the same tomorrow as today, so it makes sense to quit tomorrow instead. But where we're headed now may not be where we want to end up in the long run.

"Because every step along the way, development seems like a good idea to the person who owns the land. And then to the person who buys that house, shore protection seems like a good idea. But where that carries us is toward a coast that is almost entirely developed, and with some kind of shore protection. Then we end up losing our wetlands in most areas."

Seawalls or dikes — the ultimate shoreline armoring — also mean that, eventually, a lot of people live below sea level. "And then we're ending up in that very precarious situation, which leaves us with more than a hundred cities and towns — many more — that look like New Orleans," he said.

A floodwall can also trap the downpour from a hurricane or tropical storm on its landward side, compounding the hazard, which happened when Tropical Storm Gaston dumped 14 inches of rain and flooded much of downtown Richmond in 2004.

To avoid all this "would require some kind of social decision to not go down that road," Titus said. "It can be done by the federal, state, or local governments, but it's probably not going to be done without a certain amount of local buy-in. People in the community are going to have to be willing to see parts of it go underwater."

WHICH BRINGS ME BACK, finally, to the dream home, out there facing the Atlantic in Sandbridge. Its strip of beach, as mentioned earlier, has to be replenished every few years with imported sand. Until the latest $15 million round, which the neighborhood agreed to pay for, these infusions were mostly financed by the general public via the Army Corps of Engineers. That practice may or may not resume, but it is, in any case, not unique to Sandbridge. The Corps is shoring up a lot of teetering real estate in Virginia (and up and down the Atlantic seaboard) at enormous expense: $120 million in Virginia since 1994; $1.2 billion for the rest of the East Coast during the same period.

And back at the lot on Sandfiddler Road, there's always insurance if things get wet. Most private insurance companies have walked away from safeguarding owners from the financial liability of losing the house they could build here, or on thousands of other coastal and tidewater Virginia lots that are vulnerable to flooding. They concluded years ago that the risk is too high.

But the Federal Emergency Management Agency (FEMA) has written more than seventy-six thousand such policies on homes and businesses in Hampton Roads and the Virginia part of the Eastern Shore, with a liability exposure of more than $19 billion. FEMA, whose mission is to come to the aid of the victims of unavoidable natural disasters, was declared nearly insolvent in 2010, so Congress voted an emergency infusion for the agency. Then came Hurricane Sandy, which called forth another $51 billion in federal help, and counting.

Federal subsidies to keep those flood insurance rates low have been scaled back (for the time being), but insurance up to $250,000 is still available as an incentive to build in spite of sea level rise, as are federal mortgage guarantees and subsidies for curbs, gutters, roads, and city water and sewer hookups. Federal tax laws provide more incentives that reduce the risk, ultimately at public expense: you can write off uninsured losses, especially if you rent the property, and you can depreciate it, too.

Clay Bernick and I visited the dream neighborhood as the wind kicked up that day, under a wall of approaching storm cloud. He pointed out a spiffy, multistory pastel vacation palace: "And there's a new one, look at that thing. That's insane! And there's no beach in front of it — there's a bulkhead."

For some of us there's enough money not to care about the risk either way, especially with the incentive of subsidies and tax advantages. But as the next chapter explains, even if you can afford to "pay to stay" by the water, it's often at the expense of another public resource: Virginia's vanishing and essential natural shorelines.

Titus once titled a presentation to local officials, "The Sea Is Rising — So What's the Plan?" He likes to quote Matthew 7:26 as he PowerPoints the way forward at these kinds of meetings: "And every one that heareth these sayings of mine, and doeth them not, shall be likened unto a foolish man, which built his house upon the sand."

ENTREPRENEURS

A heavy, steel-gray fan shell, wider than the palm of my hand, is within easy reach on the sill near my desk. It's a gracefully fluted specimen of the state fossil of Virginia, a long-extinct scallop called *Chesapecten jeffersonius*, 3 million years old. But in some ways it's as representative of the near future as it is of the very distant past.

Three million years ago was the Pliocene epoch — unimaginably remote by human measure but in geologic time quite recent. Near enough, in fact, to make comparisons with our current situation more than plausible. The geography that can affect global climate wasn't much different then. The positions of the continents — they merge, split, and slide around on the skin of the planet, given enough time — were nearly the same as now. The mountains haven't worn down or reared up much since then, nor have the courses of major river systems shifted drastically enough to alter the climate.

The most immediate similarity, though, is this one: atmospheric CO_2 concentrations during the Pliocene were about 405 parts per million (ppm). Back in the late 1980s, when the geologist Harry Dowsett started looking for climate indicators in fossil shells, we were at 350 ppm. We're past 400 ppm now, and quickly adding more all the time.

So as you read this, we are already back to an approximation of the CO_2 concentrations of the Pliocene. That epoch was also, and not by coincidence, the last time Earth was as warm as it is projected to be again, around the year 2100.

"The Pliocene is in some ways an imperfect analogue for the future," Dowsett told me during a visit to his U.S. Geological Survey research lab in Reston. "One of the reasons is that the slope of growing greenhouse gas concentrations is so steep right now. I have no idea what sorts of feedbacks will result from that." The Pliocene climate was comparatively stable. Ours is changing fast, and that will continue as greenhouse gas concentrations intensify.

Pliocene sediment samples have been collected for Dowsett's research at dozens of sites around the world, most of them from the ocean floor. They give us a composite record of that ancient climate, and it is detailed enough that it is used to test, and improve, current climate models. If the models are working well, they should be able to start with known data about topography, CO_2 concentrations, sea level, and other factors, and then re-create the Pliocene climate. Where the model results differ from Dowsett's reconstruction of that ancient climate, retoolings or explanations are called for. Either the model is out of whack, or Dowsett's paleogeology is awry. This is a powerful kind of corrective feedback.

My scallop shell was dug out of a shoal of mud and sand at a place called Morgart's Beach, along the south bank of the James River near the town of Smithfield. The shell's surface striations indicate that there was less seasonality in this place then — less of a temperature difference between summer and winter. It is larger than the scallops that are found here now, which indicates warmer water.

But Dowsett and his research team look mostly at the remains of the much smaller animals we've already encountered in an earlier chapter: the microscopic, jewel-like shells of a broad class of plankton called *foraminifera* — "forams" for short. They are washed and sifted at the lab from bags of collected dirt samples. Identifying species and assaying their chemistry, the scientists can measure the depths, temperatures, and salinity of the 3-million-years-ago oceans, and the ratios of certain isotopes in their fossilized shell material indicate the extent of the icecaps. That, in turn, tells much about what Virginia and the rest of the planet looked like under the warmer, CO_2-rich canopy of the Pliocene climate.

"When we started doing this, it was all academically interesting," Dowsett said. "But at some point along the way, something kicked in that actually made it real for me. I realized it was no longer just academic. It was: This really is happening. I don't have to try to make it sound relevant. It really is. There was a sea change."

As we talked, Dowsett and his colleagues were especially wary of committing the "attribution error"—characterizing short-term phenomena as evidence of climate change—though there were ready anecdotes about how the farm pond no longer ices over, and the kids rarely see snow anymore. I asked if, during that particular summer's record run of triple-digit heat, they sometimes saw Northern Virginia through one eye and the Pliocene with the other as they commuted to work. Glaciologists or forest ecologists or marine biologists I have spoken with—their research subjects are melting, withering, or dying off during the course of their careers—often become somber when asked about the future. This group laughed.

"We're accustomed to thinking in deep time in general, and so it's easy for us not to get excited," Dowsett said. "We know about variability and blips. You're very unlikely to find people in this game who relate a heat wave in August to climate change. In fact, we're probably the opposite. We would probably say, 'nah, that's a heat wave, this stuff happens.'" Geology, as the saying goes, cycles slow, but big.

Three million years ago, global annual average temperatures were 3.5 to 5.5 degrees warmer than they are now. Sea level was 80 feet higher, give or take because that is a rough estimate, but that much water can inundate a huge swath of the landscape. Marci Robinson, one of Dowsett's research colleagues, narrows the focus to Virginia: the mid-Pliocene mean annual temperature was about 70.7 degrees on this part of the coast when the sea level was at its highest. That's hot—6.3 degrees higher than the present coastal mean annual temperature of 64.4 degrees.

DURING ONE PART of the Pliocene, the Atlantic shoreline was only a little east of, and roughly parallel to, the present-day route of Interstate 95. Its geographic signature, a ragged north-south line where buried oceanic sands and gravels give way to higher-elevation granite, is called the Chippenham-Thornburg Scarp. Geologists find traces of that old shore near the fall line—the rocky falls and rapids you've seen on the Potomac River at Great Falls, the Rappahannock at Fredericksburg, the North Anna in Hanover County, the James at Richmond, and the Appomattox River at Petersburg. Back then, you'd be standing in Richmond or Alexandria and looking out at the ocean.

No climate scientist I know of has predicted we will see anything like that much sea level rise, at least not within centuries. But the image may help us sense some of what is in store for Virginia's existing wetlands along the Bay and the coast during coming decades. With the onset of sea level rise and in the absence of planning, most of those natural areas will drown.

We depend heavily on them, indeed. Unless you're a crabber, a fisher, or an aquatic ecologist, these dun-colored, buggy, pungent patches along the shoreline may not seem important or have much appeal, but Virginia's marshes and wetlands are vital for many reasons. They are nurseries for fish, they filter pollution, reduce the impact of floods, act as storm surge buffers, and protect freshwater supplies from saltwater intrusion.

A bumper sticker that warns, "No wetlands, no seafood!" overstates the case, but not wildly. If you and I yank off our shoes and go for a stroll in fragrant marsh muck, we might reach down and pull out a handful of it for a closer look. That decaying stuff and the organisms writhing around in it are essential for the aquatic food web. Invertebrates and small fish feed on them, and they in turn are consumed by many other species including ones we eat. The tide-flats, a similar kind of habitat, supplies worms, clams, snails, and other key sources for a long list of fish and shellfish. One study lists twenty-five different species that depend on these wetlands as nurseries or for foraging, spawning, refuge, or as corridors between habitats in this region.

Because so many varieties of fish also spend much of their lives in the ocean, "significant loss or degradation of these Bay habitats could also affect the northeast U.S. continental shelf marine ecosystem," researchers have concluded. An estimated 66 percent of the mid-Atlantic commercial fisheries' catch depends on the region's beaches and coastal marshes for nursery and spawning grounds, including blue crabs, rockfish, bluefish, flounder, sea trout, mosquitofish, spot, mullet, and croaker, killifish, mummichog, perch, herring, silversides, and bay anchovy.

In Virginia, about 660 square miles of shallow water — less than 2 feet deep — make up this highly productive habitat. Given the sea level rise that most recent studies anticipate, about a quarter of it will disappear by 2050, and perhaps as much as 75 percent by the end of the century. One federal overview of research went further, calling it likely that most mid-Atlantic coast wetlands will not survive the century.

Submerged beds of eelgrass, widgeon grass, and other kinds of vegetation are also essential for juvenile fish and shellfish that use it to feed and hide from predators. These plants are vulnerable to rising temperatures as well as to drowning in too-deep waters, and they are already severely diminished by pollution, silt, and warming. The VIMS study from which many of these projections come gives strong odds that virtually all eelgrass beds could disappear by 2100, especially if they are unable to migrate inland. "Preserving . . . the Bay's essential shallow-water habitats should be a high conservation priority," the study concludes.

Some natural shorelines will be lost no matter what we do. Either they are already walled off by buildings and roads and will drown as the waters rise, or the land is too steep to allow wetlands to migrate inland, or the water will advance so quickly that it will outrun these plants' and animals' ability to expand their ranges. But we can ensure the survival of many other wetlands along with their wildlife, the fisheries they support, and their other ecosystem services such as flood control and buffering against storm surges, by making room for them.

For that kind of conservation to occur, landowners will have to yield shoreline property to the encroaching water instead of defending it. That is not happening now. Instead, VIMS tallies have found that local and state agencies approved 186 miles of armoring along tidal shores between 2000 and 2012. But when shorelines are locked up with rip-rap, bulkheads, dikes, seawalls, revetments, groins, jetties, dunes, beach replenishment, and other barriers, then natural areas drown as the waters rise in front of the new fortifications.

THE COMPLEXITIES OF shoreline real estate have been much on the mind of the developer Thomas Dingledine. He owns a home along the Bay, but more significantly, he has put eight years into plans for developing a major new resort there, on a finger of land in Northumberland County called Bluff Point.

If you have a stereotype of a short-sighted real estate developer in mind, chances are better than fair that Dingledine does not fit. "If we're going to change society, we have to transform the way business people think," he once told an interviewer, "which means changing the way they are educated. . . . Business must become more about people than about profit." He and his family are high-dollar philanthropists, funding scholarship programs and a variety of U.S. and international social action programs. He described Bluff Point as environmentally responsible, with a "triple bottom line: people, planet, and profit."

Dingledine sought a controversial special exception for a zoned conservation area that lies along a marsh — an exception that allowed for intensive development, though not in the marsh itself. His plan was ultimately approved by the county: a 900-acre spread, part of it near the Bay marshes, with 530 residences, a ninety-room boutique resort hotel, a marina, and 34,000 square feet of commercial space for shopping, dining, and a nature center.

The project was opposed by environmentalists, however, including the principals of the Conservation Institute at the College of William and Mary, the Chesapeake Bay Foundation, and others. Its southern parcel included extensive tidal marshes and wetlands, according to Karen Duhring, a coastal resource management scientist at VIMS. It is low and flood-prone, she said, and a planned marina basin and entry channel were likely to harm eelgrass and other aquatic vegetation, highly prized as habitat for Bay organisms. "We really can't afford to lose any more of those valuable and scarce resources," she told me. (VIMS's role is to advise regulatory agencies, not to oppose or support construction proposals.)

Most Virginia shorelines and wetlands are private property, Duhring warned: "For long-term sustainability of these resources, we really do depend on the private sector and private property owners to help make room for wetlands in the future." Or as one federal study helpfully suggests, "It is also a good time for all of us to ask whether this generation should continue to build new communities in vacant land vulnerable to a rising sea."

Meanwhile, we continue going backward. Piecemeal local decision making obscures the total picture: the future of many wetlands is being foreclosed as we ignore sea level rise and allow more construction and then more armor to wall it off from the water.

Of Virginia's dry land that is less than 3 feet above sea level — close, indeed, to trouble — only 7 percent is protected from development by conservation restrictions, and even those limits are not always reliable. Already built upon to some degree: 61 percent. Legally open to be carpeted with new houses, hotels, retail businesses and roads, despite the disappearance of wetlands and the onset of sea level rise: 32 percent. Virginia, as one study points out, has "few institutional limitations on coastal development."

That hornet's nest left Tom Dingledine — who sees himself as an advocate for a new standard of environmental sensitivity — facing a lot of opposition. I met with him one bright spring morning at his shoreline headquarters with a view of Bluff Point across the water. His bet, he said, was that family-friendly design has become more important than swaggering scale in these kinds of homes. He also figured that although it comes with a higher price tag, buyers would pay that premium if they were convinced that a developer's claim to green-ness was credible.

Part of his environmental monitoring included time-lapse satellite images of the disappearance of polar ice, too, and he asked his planners to estimate the impact of up to 3 feet of sea level rise on his development plan. Even the lowest elevations of the project stayed dry in that scenario, he said — but not in a storm that pushes water.

Rather than invite that future mess, many conservationists and planners argue, Bluff Point and its kindred, no matter how many "green" attributes they include, are a losing proposition and just shouldn't be built on or near the water anymore. Dingledine, looking over at an easel displaying big plat maps, vowed that he would not abandon homeowners to the flood. His stated horizon is a century out. "This is a legacy of mine, no matter what," he told me. "All these things go back and forth in my mind, as I look at what's the right thing to do, what's the best plan of development."

When the sea level rises, it will take astronomical sums to protect the future coastal construction that Virginia now allows. Just the same, if there were instead a statewide moratorium on shoreline development, Dingledine pointed out, "then immediately what you've done is just devalued property by massive amounts of money." So: big losses either way, though vital wetland areas are sacrificed in one scenario, and have at least a chance of survival in the other.

Another idea would be to levy special district taxes against the possibility of sea level rise, just as Sandbridge homeowners are now taxed to pay for sand replenishment. The money could be used to move the houses back farther from the water, or jack them up higher on stilts. Or to dismantle them entirely, preserve wetland opportunities, and then compensate the dispossessed owners. If sea level rise doesn't materialize as a threat in some happy future time, the fund dissolves and its shareholders are richer.

Dingledine was mulling that adaptation to sea level rise, which is often referred to as "rolling easements." As an Environmental Protection Agency primer explains, development might be allowed "with the conscious recognition that land will be abandoned if and when the sea rises enough to submerge it." So until the land is immediately threatened, it can be built upon. Once the water arrives in earnest, by advance agreement among all parties the land converts to wetland or beach.

This idea doesn't fit comfortably into polarized political categories. It means government can promote the orderly pursuit of capitalism by looking out for the public interest. Especially in a crisis.

Months passed. At the end of the year, Dingledine announced, to the astonishment of some observers, that a 37-acre parcel had been donated to James Madison University for a field research laboratory and perhaps a small educational facility. The balance of Bluff Point, 860 acres, was placed in a perpetual conservation easement held by the North American Land Trust.

"The world is different today and the knowledge is greater than it was." Dingledine's announcement read. "While I can influence and even control some items, I cannot control what Mother Nature may send our way in the decades to come and while such prospect has no impact on today, it may in the future and thus be a challenge to true long term generational success." That sensibility could inspire some courage in elected officials as they face the prospect of sea level rise. Alternatively, it may continue to be ignored.

CLIMATE CHANGES already under way in coastal areas are simultaneously large-scale and intricate, zoom-in and zoom-out. Both can be seen a couple of counties west and 40 miles inland from Bluff Point, along a precarious line of planks nailed onto pilings that leads out onto Sweet Hall Marsh.

This pristine, 900-acre freshwater tidal wetland is one of the largest of its kind in the United States. Privately owned for generations and kept as a hunting preserve, it is also a research lab of sorts. "These are some of the most rapidly changing marshes — this is a hot spot for sea level rise," according to Carl Hershner, director of the VIMS Center for Coastal Resources Management. "It's not like tomorrow we're going to wake up and find there are no marshes here. It's just that these marshes over the next hundred years or so are likely to change dramatically in character."

The marsh straddles a big ox-bow bend of Virginia's lethargic Pamunkey River, about 50 miles upstream from the Chesapeake Bay and just above the point where brackish water yields to fresh. "When we started working out here thirty years ago, you would have been in the midst of a stand of *Spartina cynosuroides* — big cordgrass — because the marsh was dominated by that," Hershner said.

He pointed to spots where duck hunters would use the walls of *Spartina* along the river as blinds. These days, it only grows in a few patches in slightly higher elevations, replaced mostly by a lush, low-growing, more water-tolerant plant once known as "tuckahoe" and now called arrow arum. "In that period of time, sea level has risen probably 5 inches in this area," Hershner said. As sea rise accelerates, incoming sediment is no longer able to keep pace, so the waters deepen. This marsh will eventually become a tidal mudflat without vegetation, and then a shallow pond.

As the water rises everywhere along the coast, adaptive management will be essential for natural areas, Hershner said: cautious decisions, continuous monitoring, revisiting decisions, informing and motivating policy makers.

Sometimes they just aren't listening. "That's why we describe ourselves, after decades of fighting this, as functioning cynics," Hershner said. "We've been beaten back enough to know what's coming, but we're still able to get up and continue to fight the fight, because we've got to do it. . . . We now have enough insight to see how things are changing, and the public's motivation to deal with them will probably never be greater. It is really within our capacity as a society to say, 'Yup, now's the time.'"

The combination of real estate development and sea level rise is affecting many species already, including the horseshoe crab, a mesmerizing, many-legged animal you or your kids may have played with in the "touch tank" at a nature museum. Horseshoe crabs are not really crabs — they are closer kin to spiders and scorpions.

I clambered into a skiff on a rainy evening with the Virginia Tech fisheries biologist Eric Hallerman and his graduate students, who were there to survey horseshoe crab populations. We were on isolated, somnolent Tom's Cove off Assateague Island, part of Virginia's section of the Delmarva Peninsula. A full moon loomed behind broken clouds, and the horseshoes made their ponderous way shoreward to spawn. Colliding gently in the shallows, they clicked like billiard balls. Then their big curved shells moved onto the beach like tanks in a science-fiction army.

Horseshoes usually mate on sandbars and beaches, and that is where the females deposit their eggs. Those spawning grounds are diminishing because of coastal construction and sea level rise, but the horseshoes will abide, even if in sharply reduced numbers. "I'm not particularly worried about them," Hallerman told me. The species has, after all, persisted through 200 to 300 million years, and many changes of climate and sea level.

So, amid the calamity of sea level rise, who would care if their numbers taper off? Their situation is not isolated from the rest of shore life, though. Many other species depend on horseshoes. The greenish egg masses left in the sand by these fecund animals — each female deposits about ninety thousand eggs per season — are a major source of protein for nearly a million shorebirds migrating through the region each spring.

"Two dozen species of shorebirds that we know of feed heavily on horseshoe crab eggs, and a few are already imperiled," Hallerman said. "The eggs are a superabundant and energy-packed resource. Marine turtles prey on horseshoe crabs a lot, so a loss of them might be a significant hit for those turtles."

Birds are also at risk because they use several different habitats in what are now 2,100 square miles of tidal wetlands in the mid-Atlantic region. These include marshes with varying levels of salinity, swamps, rocky and riverine and muddy tideflats — the wetlands that are not likely to survive even 3 feet of sea level rise.

The Chesapeake, with its low and vulnerable shoreline, provides habitat for canvasback, mallard, redhead, and American black ducks, tundra swans, geese, herons, snowy egrets, and many species of songbirds, as well as several kinds of increasingly rare sparrows, black rails, Forster's terns, gull-billed terns, black skimmers, oystercatchers, red knots, and piping plovers.

On the "winners as well as losers" account books, it is apparent that losses of biological diversity will markedly outweigh gains. "Sharp declines in populations and local extinctions are certainly possible in the mid-Atlantic region," the U.S. Geological Survey wildlife biologist Michael Erwin told me, though some wintering waterfowl would probably be able to take advantage of more open water.

"It's potentially going to affect resources for a good part of the hemisphere," Erwin adds. "If a lot of the mudflats become unavailable or less productive all up and down the Virginia coast, then something like thirty species of migratory shorebirds will be affected, and several of those have already been listed as candidates for threatened or endangered status."

Erwin has studied birds along the Atlantic since 1968, as many have become imperiled. He wonders whether, as sea level rise advances, our own species will be moved to take measures to protect them. "I would say it's unlikely. Almost any wildlife resource, I'm afraid, would be way down on the totem pole compared to all the things that humans hold near and dear: the economic things, for instance."

That burgeoning conflict between protecting biological diversity and meeting our shorter-term demands will sharpen as sea level rise becomes a destructive force, Hallerman believes. "It doesn't even have to get that desperate. I'm really worried that in our self-absorption we're going to drive lots of populations to extinction, and even species to extinction. It's one of those things I worry about on nights when I can't sleep."

HALLERMAN'S HORSESHOE census at Tom's Cove was on the southern end of Assateague, a 37-mile-long, mostly wild barrier island, a National Seashore that hooks around Chincoteague Island and its namesake town and wildlife refuge. Hundreds of thousands come each season to see the famous Chincoteague ponies and the clouds of songbirds and shorebirds and, especially, to enjoy the beach. In the fifty years since this National Seashore was created, the town's economy has moved steadily from fishing to tourism, and currently reaps about $42 million a year from 1.3 million annual visitors. For the region, the figure is $150 million in tourism revenue.

The former manager of the Chincoteague National Wildlife Refuge, Louis Hinds, arrived on the scene a few years ago. His inaugural public meeting began around a big oval table and the whole agenda — oddly, he thought at the time — was a discussion of parking lots on the beach. The mayor of Chincoteague was there along with a contingent of townsfolk, and representatives of the National Park Service, which administers the National Seashore.

It soon became obvious why the one-thousand-car parking lot created the sharp tension Hinds sensed in the room. It keeps getting washed away. So does the causeway leading across a big artificial pond to the recreational beach and the parking area. They are increasingly expensive to replace, and, in any case, the beach itself is eroding away.

Much of the town figures that if the parking lot disappears in the surf, or if it were moved to another beach a mile or so north, the tourists would vanish too. But replenishing the sand barrier would cost $24 million, and that would only endure for two to seven years before more millions would be required.

The state of our federal budget is plain enough — both the National Park Service and the U.S. Fish and Wildlife Service have endured many years of stringent economies. A proposal to levy a hotel tax in Chincoteague to pay for keeping the parking lot in repair withered for lack of local support.

The costly parking lot rescue has been debated in tumultuous meetings, sharp-fanged blog posts, and a Washington hearing during which local congressional representatives who are budget hawks on most other occasions railed at federal administrators, charging them with abandoning the town's economy. What do we owe each other in these cases, and how can decisions be made on a better basis than who can show the fiercest lobbying face?

As it happens, the controversial parking lot also sat in some of the best endangered-species habitat on the wildlife refuge. The whole of Hinds's domain is also a parking lot for more than three hundred species of birds such as snow geese, black skimmers, loons, and great egrets, partly attracted by

2,650 acres of freshwater impoundments. These, he predicts, will probably disappear by midcentury as sea level rise advances on the slender refuge.

We got into his pickup truck to reconnoiter the southern pasturage of the famous wild Chincoteague ponies — descendants, it is said, of escapees from a Spanish shipwreck of three centuries ago. The view was framed by the skeletal remains of clumps of dead pines. Their roots have been flooded by salt water that is probably pushed up by sea level rise. From the same cause, the grass here is already too poor to support the pony herd year-round, and their forage dwindles. The scene is likely to be underwater in a few decades.

"We're tied to our traditions, and that's good," Hinds said. "Traditions make us strong. They let us know what's happening around us. They give us comfort. But we also have to plan for the future. Obviously, this landscape is already changing."

We drove out on the long, storm-worn beach, through tumbling air that rocked the truck. The shoreline curves around to the west along a debris field with a view of Wallops Island, all of which is federal land occupied by high-tech military training facilities and NASA rocketry. It's a curious juxtaposition, and one that Hinds finds instructive.

"We're looking at a barrier island that is filled with some of the most technologically advanced industries that this nation has," he said. The noisy, crash-prone $4 billion Osprey hybrid airplane/helicopter thumped the air over the channel that day. The superstructures of a mock-up aircraft carrier, cruiser, and destroyer were silhouetted on the horizon. They are used to train military men and women for advanced monitoring and antimissile operations. There are facilities to stage air attacks simulated with drones, and the NASA/Virginia spaceport that supplies the space station is also over there somewhere.

Sea level rise is moving in on Wallops Island, too. "Their response is going to be different than ours will be for the natural environment," Hinds said. "They've just finished a $34 million stabilization and protection project, where they put in a long rock wall and then built a beach over it and in front of it. From a national perspective that's a pretty damn good insurance policy for your billion-dollar investment."

But how about $24 million for a one-mile, temporary recreational beach to support a temporary parking lot? That may not be such a good investment, though if someone doesn't make it, citizens of Chincoteague who have done nothing in particular to deserve the economic hit may well suffer. The problem with that investment is that the next bill, even higher, will come due soon.

"That's the kind of decision communities and eventually political leaders will have to make. Where does it make sense, where doesn't it?" Hinds said. "We've got both extremes, and we can see them sitting right here in this pickup. Sometimes protection is right, but sometimes moving provides resiliency. It's planning — it's what we should be doing as a living, thinking species: looking forward to the future.

"Until this country comes to grips with catastrophes like Katrina and New Orleans, or Sandy and the New Jersey coastline, until we put this up on the national stage and say how we are going to manage it, we're just going to be kicking this can down the road for our sons and daughters and grandchildren to handle. We need to be smart, deal with it now, give guidance and direction on a national level."

The don't-hide spirit Hinds favors will take hold, sooner or later. You might think of it as a variation on the entrepreneurism of Carl Hershner as he and VIMS try to slow down the destructive process of hardening shorelines, or of Skip Stiles, who tries to educate local officials about sea level rise, wetlands, and new construction.

Or of Tom Dingledine, as he puzzled out how to reconcile climate disruption with a new real estate venture and, ultimately, relinquished an outmoded dream in exchange for creative adaptation. That outlook will be indispensable as the atmosphere of the Pliocene, and its retreating shorelines, arrives.

RESETTLEMENTS

Mosquitos matter. Virginia has fifty-seven different kinds of them — "each one as different in habitats and behaviors as a hawk and a hummingbird," as one expert says — and about ten can carry a variety of infectious diseases. Out in my backyard, and probably in yours if it's a little shady, we now have robust populations of those incredibly annoying black-and-white-striped Asian tiger mosquitos, inadvertently imported and then established in Virginia and the rest of the Southeast by around 1992. Their active seasons are lengthening even in the Shenandoah Valley, where severe winter cold used to limit their reproduction.

Potent biters, tiger mosquitos and their many cousins in the insect kingdom are usually a nuisance but not a tragedy. That can be circumstantial, however. It depends on chance, culture, and climate. I was reminded of that a few years ago when I came across an odd book. It previews some of the climate change you're familiar with if you've read this far, but with an eye toward Baby Boomers making plans for retirement.

You might, the book counsels, be an "ecological speculator," figuring out how to profit from the trends set in motion by climate change, or a "snowbird" who would follow the altered seasons from one optimal perch to the next, or a beachwalker, a "sustainable hedonist," or a "sun-seeking road warrior." You could decide to move nimbly from place to place to stay out of the way of climate troubles, or just choose a single calm spot where you could roost, untroubled.

Well, that's my secret, somewhat sheepish climate plan at times, too. Get me and my family out of Dodge, so to speak. It's what Americans have had to do when environmental or economic disaster strikes, and sticks: we migrate. But the book's chipper take on planning leaves a few important things mostly out of account. Mosquitos, for one. They will matter even more as the climate heats.

And all that beachy snowbird hedonist music is for soloists and duos, and affluent ones at that. For most Virginia folks, our quality of life and sometimes our survival will depend on being part of an orchestra. That is reflected in the dark humor I sometimes hear from scientists around the state who contemplate climate disruption. "You don't have to worry about trying to protect that forest from more suburban sprawl," a biologist told an ecologist. "Roving bands of survivors will cut it all down for firewood when global warming comes and anarchy breaks out."

Well short of that desperate scenario, it's worth wondering what may befall our support systems, our communities — and even the fortunate itinerants coyly labeled above — if more frequent climate disasters are on the program. They will send waves of displaced Virginians and other Americans anywhere that seems to offer safety and subsistence. Whether we are the refugees or the refuge givers, the openness and resilience of our social networks will be tested. Our safety nets may prove to be supple, or they may be a descending series of cobwebs. Our recent national history gives uncertain signs of both possibilities.

Indeed, this field of research — migration studies — is said to be galvanized, given the likelihood of climate disruption. Across the United States, internal migration will be the foremost option, one study predicts, to escape heat, droughts, sea level rise, floods, and extreme weather events, particularly along the Atlantic and Gulf coasts.

As you'd expect, different kinds of environmental disasters produce different responses. Some are preceded by warning signs, but we're blindsided by others. Some endure, others pass quickly (though never quickly enough!). We may be able to adapt to slow-onset, cumulative disasters such as drought, deforestation, or soil degradation, if they are gradual and the environment and our community bonds can accommodate them. That is to say, if food distribution, education, security, health, and economic support networks hold.

Rapid-onset disasters — think hurricanes, tornados, floods, wildfires — may produce a quick evacuation and return. But victims may be compelled to migrate for longer terms, or permanently. Ethnic-minority, low-income, low-education, and female-headed households are less able to return to their homes, research has confirmed.

The studies often draw from our two preeminent environmental catastrophes, the Dust Bowl — a slowly arriving, long-duration example — and Hurricane Katrina, which, though brief, caused a diaspora of fleeing Americans that, according to some sources, rivaled the Dust Bowl.

The Dust Bowl crisis pushed waves of refugees from Texas, Oklahoma, and Arkansas mostly westward over several years in the later 1930s. They fled ever more defeating conditions of successive drought, failed crops, and killing dust storms that left dwindling possibilities for adaptation-in-place.

The arrival of "Okies" in the western states, principally California, incited public hysteria, historians tell us. A "sea of antagonism" awaited the refugees. During one bleak period in 1936, Los Angeles police were stationed on the Oregon, Nevada, and Arizona borders to roust, search, and turn back the tide of victims of environmental catastrophe. It was called a "bum blockade."

Public services of all kinds, especially public health systems, were strained, sometimes past the breaking point, for several years. As was goodwill. "California is faced with economic chaos and financial ruin through the influx of thousands of families, displaced in other states and pauperized by the Depression," a manifesto by California business and agricultural leaders stated. It argued that the cause of the problem was the state's too-generous relief programs.

Hurricane Katrina's aftermath tells a more recent migration story. "Two weeks after it blew through the U.S. Gulf Coast, it's clear that Hurricane Katrina has resulted in the largest displacement of Americans in 150 years — if not the largest ever," one social scientist told a reporter. "The scale is monumental. It's as if the entire Dust Bowl migration occurred in 14 days." Another said, "This is the biggest resettlement in American history. A whole city has been uprooted."

The official count was that 3,300 Katrina refugees fled to the Richmond and Virginia Beach–Norfolk–Newport News areas alone. Generosity in those and other communities around the state was inspiring. That our schools and churches, volunteer organizations, and local governments would respond similarly to the next flood of refugees is beyond question.

An open question or two does remain, however: What would happen as successive new surges of victims arrived? And then, what if months and years pass and the refugees are still around?

The research gives us more than anecdotal evidence of how help is sometimes offered. "Before Katrina, one town (Lonsdale) in rural Arkansas had fewer than one thousand residents and a single African American," the political scientist Daniel Hopkins of Georgetown University has written. "According to a state official, the Ku Klux Klan was active nearby. Yet for a short while, Hurricane Katrina changed those demographics dramatically. Overnight, a local church camp became a shelter to 350 evacuees, almost all of whom were poor African Americans."

One community leader told Hopkins that "the majority of the people here . . . were very angry with us. In the beginning, people were very prejudiced. Once these people got to know them, everyone's outlook totally changed." One church leader was told by an evacuee, "We didn't know white people could love us." In the days after the storm, residents of many host communities, including those in Virginia, mounted a response that "undermines simplistic notions of racial threat," Hopkins found.

The narratives don't always turn in that direction, however. "Large-scale migrations transport place-specific disasters to other communities," as one study puts it. Many communities receiving evacuees from Hurricane Katrina differed culturally, racially, and economically from the Gulf Coast. Evidence of discrimination was found in housing placements, government and charitable programs, and from private citizens offering help. Illegal immigrants were eligible for disaster aid but not immunity from deportation, so many were fearful and did not seek assistance.

Hopkins compared Houston and Baton Rouge, which both absorbed huge numbers of Katrina victims. Houston took in 250,000. A year later, 150,000 were still there. "I became interested in where the evacuees were going and how they were being received," Hopkins told me. As time passed, attitudes began to harden. Sympathy subsided, animosities built. Images of people trapped on rooftops in New Orleans' Ninth Ward lingered, but after a couple of months it was more often heard, Hopkins said, that "it's time for the evacuees to pick up and move on."

Crime rose in both cities, and in Houston it became a focal point for public concern. "It was one of the ways that locals understood the evacuees as a group. Even when I interviewed individuals who were closely engaged in providing services to the evacuees, they would nonetheless tell me about how when their car was broken into, they suspected it was Katrina-related."

But even the year after Katrina, both the number of violent crimes and the news coverage of crimes in Houston were actually *down* from their peaks in 2002. In 2002, surveys found that only 13 percent of citizens perceived crime as Houston's most important problem. Nonetheless, in 2006 — the year after Katrina refugees arrived — that crime-is-the-biggest-problem number rose to 31 percent. "After the arrival of the evacuees, Houston residents were paying more attention to fewer crimes," Hopkins wrote, and Katrina evacuees were blamed.

There were no separate surveys of the attitudes of Houston's black community, only anecdotal data. "But it was one of the things I was really struck by," Hopkins told me. "People said that on the one hand the response in the Houston African American community had many features in common with the white community, only more so. They mobilized in a very powerful way to help." But some also worried that they might be judged by the perceived misdeeds of the evacuees. "So I certainly heard black-oriented radio stations in Houston where some of the anti-evacuee sentiment was as strong as, or stronger than, it might have been in other parts of the city," Hopkins said.

According to the FBI, the increase in violent crime in post-Katrina Baton Rouge, compared with two years earlier, was 20 percent — much steeper than in Houston. Sentiments toward Katrina evacuees turned negative in Louisiana, too, but "the conversation around the evacuees focused more on aid, on welfare-type benefits and FEMA assistance," Hopkins said. In other words, local residents turned to familiar cultural and political story lines, to explain refugees displaced by a natural disaster.

"Clearly they suffered for reasons that were out of their control, but they nonetheless were targets of some blame, as well as a significant amount of sympathy. The community mobilization to assist the evacuees in both places, in all of those places, was tremendous," Hopkins found.

The mythologies within which refugees are viewed, then, help determine how they will be treated. For example, are we in Virginia all "downstream" together, in terms of coping with climate change? What will we feel we owe each other, if calamity displaces whole populations?

The sociologist Elizabeth Fussell has written that perhaps scientists and policy makers can identify policies that will allow residents of disaster-affected areas to remain in their home communities, or smooth their incorporation into new destinations. She has also concluded that where possible, it is better to disperse migrations so that their impacts do not fall squarely on a few localities. Volunteer efforts provide on a heroic scale, but cannot be expected to do so past the short term. Hopkins found that government leadership and reliable financial and other assistance inspires and strengthens effective local campaigns.

The most recent federal climate assessment predicts that, "in the future (as in the past), the impacts of climate change are likely to fall disproportionately on the disadvantaged. . . . The fate of the poor can be permanent dislocation, leading to the loss of social relationships and community support networks provided by schools, churches, and neighborhoods."

That is already part of the climate conversation in Norfolk. Cyndi Simpson, minister of the frequently flooded Unitarian church whose congregation is grappling with whether to move to higher ground, sees their anguish but also their good fortune: at least they are able to relocate.

"I have grave concerns here in Norfolk about the social justice aspects of continued flooding, though," she says. "If there are neighborhoods that get assistance and ones that don't, how will that be decided? If there are folks that get help, and folks that don't, why is that going to be?

"As a citizen of Virginia and of Norfolk, I can tell you that anyone who thinks that race is not going to be an issue, in a bad way, is out of their mind. So I worry about people who are already disenfranchised for whatever reason — poverty, ethnicity, bias. How is that going to play out?"

MIGRATION WILL NOT be limited to one species but will expand the ranges of infectious diseases and their carriers such as mosquitos and ticks. Malaria, Zika virus, dengue fever, Chagas disease, and chikungunya virus, for instance, are on the list of potential threats. They can have seriously debilitating, sometimes lifelong effects on victims and broadly damaging social impacts. Zika virus can cause severe head and brain malformations in human embryos. Some of these diseases can be fatal. How well our communities cohere in an epidemic will matter as much as the heat, the rains, the bugs, and the disease.

The research linking climate change and disease is unsettled. Early on in discussions about places like Virginia, the disease ecologist Richard Ostfeld told me, this was the thinking: higher temperatures mean tropical diseases will expand their ranges north. Soon, however, "there was this counterpunch by people who said, 'No, that's all wrong. Our models suggest that at most, these diseases are going to shift in their distribution with no net increase.'"

Diseases like dengue fever, in other words, might move higher in latitude, but then shrink equally at the southern end of their range as the heat rises (though that may be little comfort for those of us living on the northern side of the picture).

Lyme disease, which spreads via ticks, is an increasingly serious problem. Models had forecast a climate link, but Lyme disease's expanding range is not clearly related to climate change. Nonetheless, reported cases quadrupled in Virginia over the first decade of this century to 1,000 a year, and the Centers for Disease Control and Prevention now says they are underreported: a more accurate number of infections would probably be 10,000. If that is correct, then a major fraction, some 1,000 to 1,500 new victims each year, will suffer long-term, recurring symptoms of fatigue, musculoskeletal pain, and troubles with cognition. Climate-related or not, Lyme disease illustrates that infectious disease can travel, and settle in.

The battle over how serious a problem climate change will be in promoting disease continues to rage, Ostfeld said. "This has been one of those topics in science where people carve out extreme and in my view indefensible positions as we ever so slowly converge on the truth, which is somewhere in between." He occupies a cautious middle ground. "There will be a net expansion for many of these diseases," he said, but the degree of that expansion depends very strongly on how well the human species can pull together to defend itself.

The Southeast is a danger zone. Malaria, for example, is always tapping on the door and hoping to be allowed in. Virginia's state health service reports about 60 new imported cases of malaria here each year. "The question is, are those clusters of imported cases going to ever result in the establishment of a new endemic site for these diseases where they've been wiped out for decades?" Ostfeld said.

Interestingly, malaria didn't get here from some tropical rainforest. It was likely one of those gifts to the New World brought to Virginia by the Jamestown colonists — many of them from deeply malarial areas of England — in 1607 or shortly thereafter. The late Andrew Spielman, a malaria researcher at the Harvard School of Public Health, told the author Charles Mann that "in theory, one person could have established the parasite in the entire continent."

The exchange only calls for inadvertent partners. "It's a bit like throwing darts," Spielman said. "Bring enough sick people in contact with enough mosquitos in suitable conditions, and sooner or later you'll hit the bull's-eye — you'll establish malaria." And Virginia, he concluded, was the most likely first landing.

Malaria was vanquished in Virginia and the rest of the United States only in the late 1940s, through an initiative spearheaded by a federal program that became the Centers for Disease Control and Prevention (CDC). I asked Ben Beard, an infectious disease specialist at the CDC, what keeps diseases such as malaria and dengue fever and the rest of the trouble away from Virginia now. The answers: federal and state health campaigns, window screens, air-conditioning (it keeps people indoors), tighter home construction, efficient drainage techniques, and the spread of somewhat mosquito-hostile hardened urban areas. "Other countries have open sewage, no clean water, no window screens, no basic health infrastructure," Beard said.

That can also describe areas hit hard by climate-related disasters and/ or economic problems. Where people are in trouble, disease can thrive. In suburban Kern County, California, hundreds of swimming pools went untended and bred green algae and clouds of mosquitos when the recent Great Recession arrived. "The likely reasons for neglected pools are the adjustable rate mortgage and associated housing crises in Kern County and throughout California, which have led to increased house sales, notices of delinquency of payment, declarations of bankruptcy and home abandonment," a subsequent research paper summarized. Kern County suffered a 300 percent increase in those notices of delinquency.

The result was an epidemic of West Nile virus, which causes a kind of encephalitis that can inflict serious symptoms and, rarely, fatalities. (It is now a permanent part of the viral landscape in Virginia, too.)

"I think the climate change issue is really important. We need to be looking at it," Beard said. "Change itself is stressful. We have a public health system, but it's just sort of right on the edge. They do a great job, but our colleagues at the state health departments would say they're very concerned about what they're able to do at a local level. We're concerned about sustained funding for our public health as a nation, including for vector-borne disease programs."

As Virginia's climate comes to resemble northern Florida's, infectious disease need not be an uncontrolled hazard, Beard said. Floridians are okay, to the extent that they have a strong public health infrastructure. But he cited other states — North Carolina is one — that can no longer claim a high level of protection against vector-borne diseases because of budget cuts, a shaky economy, and shifting attitudes about the role of government.

VIRGINIA'S STATE PUBLIC HEALTH entomologist, David Gaines, keeps a weather eye on insect populations, the diseases they carry, and the likelihood that as climate alters, so does the disease landscape. For example, a family in Afton, Virginia, contacted state health authorities not long ago when they awoke to find so-called "kissing bugs" feeding on their faces. We don't pay much attention to those insects now, but according to some experts, one concern for Virginia's indefinite future will be Chagas disease. It is carried by certain species of kissing bugs, and when victims try to brush them off their faces, the insect's feces get smeared into the wound. The resulting disease generates a long list of ugly symptoms in its acute phase, grading to severe chronic heart and intestinal deterioration.

A tightly knit surveillance system and quick medical responses should be able to contain it if it arrives in Virginia. In any case, the Afton bugs weren't the kind that carry the disease, and Gaines thinks it unlikely that climate change will bring Chagas disease here. (Other infectious disease experts rate the chances higher.)

Two possibilities that are more visibly on the horizon, Gaines told me, are chikungunya virus (I know: that name sounds like satire) and dengue virus. These diseases can be carried by tiger mosquitos if introduced. Chikungunya is rarely fatal, but it can be debilitating and its arthritis-like symptoms can endure for years. They strike primarily at the wrists, hands, feet or ankles. "The longer the warm season, the higher the risk will be. In the future, we might have mosquitos that are active all year, and more chances that something could get started," Gaines told me.

His view is that our current medical care system, in which primary care doctors are a first line of surveillance and defense, invites trouble as the climate shifts. "Laboratory tests are expensive, and doctors are not even going to think about testing for something like dengue or chikungunya in patients who haven't traveled overseas," Gaines said.

But if the disease is introduced here by returning travelers, local tiger mosquitos can spread it among the rest of us. A delay in detecting its presence gives that hazard better odds, Gaines said. I asked three primary-care doctors whether his assessment seemed right, and I half-expected an indignant denial. Instead, they affirmed Gaines's view without much hesitation.

"If you have climate change, you get new environments hospitable to new vectors," he explained. "But never mind climate change. We are already vulnerable. One of those diseases could take off, particularly chikungunya, under the right circumstances. If you initially have enough of a group of infected people, that's when things can start to build. We need a health care system that is affordable, and that can identify diseases before they cause an outbreak." In that way, the fates of immigrant insects and migrating people entwine: unless we build a strong and resilient social support network, human casualties will multiply as climate change intensifies exposure to disease.

"Luck is the residue of design"— a worthwhile aphorism often attributed to John Milton—can be translated as, "Earn your luck." Much of our ill fortune, though — disease, job loss, floods, hurricanes — is unearned. As the climate roulette wheel accelerates and its bad-luck sectors expand, the odds are that more of us will fall into trouble. We may instead incline toward Samuel Johnson's competing wisdom: "A decent provision for the poor is the true test of civilization."

At this point, we tempt each other to climb onto overfamiliar soapboxes. The lucky sperm gang assails the bleeding-heart idealists; the it-takes-a-village clan rails at the self-mades. Then those rickety platforms tend to pop their rusty nails and collapse in splinters, because this brand of problem solving seldom does so.

Wiser, perhaps, to seek out common ground when possible. The climate journalist Andrew Revkin, weighing the prospects for action on global warming, has written that some of us are fundamentally communitarian, "people who want to give a group hug and see common solutions to a single problem." Others are individualists who say, "Don't tell me what to do, the best thing for us to prosper is for us all to be free, as long as we're not stomping on each other's feet to pursue our dreams." But there is also a widely shared, commonsense span of agreement on conservation, energy, innovation, and cutting waste, Revkin supposes. Focusing on those kinds of issues, you find a much larger constituency.

I vote that we add to that list of consensus priorities. If climate disruption causes migrations of people and diseases, here's hoping Virginia doesn't hesitate. We will need deep reservoirs of mutual support: help for the climate refugees among us, and a muscular public health system for all of us.

VIRGINIA CLIMATE FEVER: R$_x$

The promise of slogans — often though not always betrayed — is that a few mere words can animate us and alter behavior, if not history. "Just say no." "Can we all get along?" Here's a peppy mantra — not my coinage — that I hope carries a lot of consciousness-raising voltage in our near future: "Climate-ready and climate-friendly."

It captures two broad strategies. "Climate-ready" means planning for impacts that we can no longer avoid, such as sea level rise, dying forests, heat, and extreme weather. The "climate-friendly" part is that Virginians and the rest of the world can take steps now to throttle down our greenhouse gas emissions (fig. 23), and that will soften the coming impact of climate disruption.

It also means we need, urgently, to reverse the history of environmental ill-usage that already pauperizes Virginia's once-richer forests, rivers, the Bay, and the Atlantic. Its synergies will make climate disruption, and our own losses and survival pressures, far worse.

I have a good friend who takes the practical-seeming view that the climate feels okay right now — pretty much the usual mix of balmy and less-than-nice days — so why not just wait until things become obviously unpleasant and then talk about expensive measures? "If it ain't broke . . ." etc.

Except that climate dynamics dictate unforgivingly: the longer we wait, the more those long-enduring greenhouse gases accumulate and the more severe their impacts become. Because of the lag time between emissions and impacts, we will not experience some of the results of our current level of emissions until later this century. But when that warming arrives, it will be with us for generations.

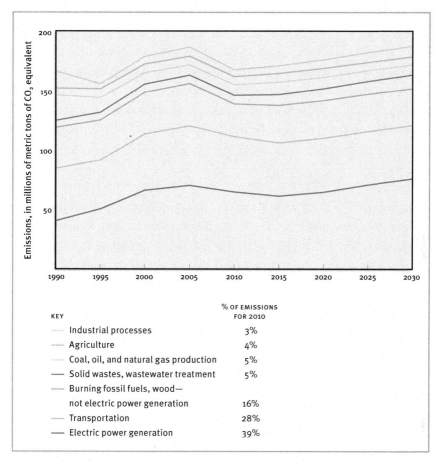

FIG. 23. Where do Virginia's greenhouse gases come from? (With projections to 2030)

Further delay also means that we are inevitably more committed to, and dependent on, the damaging carbon-based economy that now enfolds us, and we will have ever more difficulty reversing course. Acting in the near term makes the necessary retooling easier to cope with, though "easy" is not quite precise. This will be long-term, arduous, and expensive.

There are many places to find serious thought about what we have to do on the "climate-friendly" side to ward off the most serious climate disruption. The rescue plans differ, but many converge. Here's a boiled-down example: We have to reduce our greenhouse gas emissions by 50 percent in the next five years. Then we have to achieve "net zero" emissions by fifteen years after that with even more strenuous conservation and efficiency measures, and the help of engineering innovations that will collect and absorb carbon, and store it or dispose of it. That will offset the remaining, unavoidable, emissions that our civilization will still produce.

If everybody does that, we have a shot at slowing down climate change. If it works, we could make warming level off after several decades, and then keep greenhouse gases in the atmosphere at what we hope is a comparatively safe level.

You go first... No? Because the "plan" sounds demented on several counts, right? To start with, its scale. Climate disruption is your individual problem, and then again, it's not. But Americans are often invited to respond to environmental problems in this way, as if they were solely a personal responsibility: the argument is that if we all just decide to buy hybrid cars or recycle our cans and perform a long list of other tidy household measures, the world will be transformed.

Promoting this truism can have a curious side effect. It tends to paralyze discussion of responses that our government must organize — the only real hope for dealing with problems of this magnitude. Further, it trivializes by turning us all into hypocrites: how can this or that leader talk about limiting CO_2 when she flies around on an airplane and lives in a big house?

Individual consumer choices are unquestionably a valuable part of any climate solution. They cannot be the whole answer. I stand with former vice president Dick Cheney on this one: "Conservation may be a sign of personal virtue," he once said, "but it is not a sufficient basis for a sound, comprehensive energy policy." Same with the rest of the climate agenda. Individual efforts to reduce our dependence on fossil fuels indeed have the potential to help diminish greenhouse gas emissions. But they can't suffice when government incentives are powerfully aligned to make it easier and cheaper for both citizens and corporations to consume, pollute, and waste.

There's a rather obvious fairness issue, too. If you are pretty sure that most people won't be going along with the plan for the wrenching energy and emissions cutbacks outlined above, then your motivation will suffer, to say the least. The climate problem won't be resolved, and you'll pay dearly.

We can compete on who is more "sustainable." But let's make sure that everyone who contributes to the problem gets the incentives and disincentives we have coming. There's a far better chance of equitable sacrifice when we're all in the game together. Don't begrudge me my huge house. I won't grieve your SUV. Just carbon-tax us both, the way we tax cigarettes. Our behavior will adjust.

This idea links to what economists call the "tragedy of the commons." When we all own a resource in common — the atmosphere, the rivers, the ocean, the wetlands, and other public lands — businesses and individuals can exploit it for profit. Drawdown and damage to that shared resource doesn't appear on profit-and-loss statements as an expense to be made good by the user. It is instead "externalized." It shows up in consequences we all suffer collectively.

In Virginia, we have demonstrated that it's cheap — and profitable in the short term — to clear-cut forests on public land, overfish the waters, dump acid rain on mountain forests, and shove construction silt and farm waste into ailing streams and rivers, sending it downstream to pollute the Bay and enlarge its "dead zones."

At last count, for example, a quarter of the state's river and stream miles and about 80 percent of the surface area of our lakes and estuaries (including the Bay) are now "impaired" by bacterial or chemical pollution, according to the state Department of Environmental Quality. (The count is incomplete, however: 65 percent of rivers and streams have not even been assessed.)

Just so with that great commons, the atmosphere: it seems cheaper to destabilize the climate by sending greenhouse gases aloft, treating the sky as a limitless open sewer. But we all suffer the consequences, the social cost of carbon pollution, collectively. So we have to support collective intervention — through our state and federal governments — to come to grips with the problem.

The "scale" argument in favor of doing nothing is offered here at the higher end, too, but less persuasively this time. Some object to America — or Virginia — embarking on a serious climate change program because other states or nations can't be counted on to match our efforts. That logic discounts some huge benefits that favor breaking free of the status quo.

At some point, especially given Americans' special status as among the planet's most affluent, aggressive, and long-term climate polluters, we need to set a strong, hopeful example, or at least keep pace with our European allies in regulating greenhouse gases. We can't effectively inspire, exhort, or exert leverage on the rest of the world from a low moral plane.

We can also reap the economic advantages of gaining control as a nation — no matter how imperfect — over our fate. Measures to limit pollution such as the Clean Air Act, the Clean Water Act, and the federal anti-pollution and fuel efficiency standards for vehicles have always been decried by the affected industries as expensive job- and economy-killers. None of them turned out to be — quite the reverse.

THE FEDERAL GOVERNMENT has already responded to the climate issue in a limited way, and several highly visible citizens' groups in Virginia are pushing congressional representatives and the rest of government toward a range of initiatives. That is patient and essential labor — the feds and the state need every kind of prodding. Glance at these three recommendations:

- Require all new cars to achieve an average of at least 40 miles per gallon of gasoline.
- Stop global deforestation by planting at least as many trees as are felled. Trees absorb CO_2 during photosynthesis.
- Impose fees on coal, oil, and natural gas to provide economic incentives to shift away from the use of fossil fuels.

Those measures were pushed by federal agencies. In 1989. We are still 15 miles per gallon short of even the trifling gas consumption goal. The other two recommendations have not been fully implemented, either. As of this writing, our national administration talks urgently of global warming but eagerly promotes plans for new drilling, fracking, mining, and pipelining. Its actions are far too often mired in carbon dependence.

A single state cannot do this job in isolation, but state-level climate measures create an opportunity to change behavior through both education and regulation. State action can catalyze movement on the national scene, too.

Ask Californians, whose laws have spurred the development of more fuel-efficient cars for the whole country.

Today, Virginia has no demonstration acreages to simulate climate stress on crops; experiment with new strains of corn, swine, or soybeans; or find alternatives for the cattle farming that produces a major fraction of greenhouse gases. No research about climate change is under way at the state Department of Agriculture and Consumer Services, and there is none at our two land-grant universities, Virginia Tech and Virginia State, or at the state's Agricultural Extension Service.

There's no guidance from the Department of Transportation about sea level rise for state highway engineers, and none for local civil engineers planning sewer lines, roads, or other infrastructure for Virginia cities and counties, according to a Department of Transportation spokesperson.

We don't lack for examples among other states. North Carolina and Maryland already have coastal plans, however embryonic, to cope with sea level rise. Virginia does not. The state's *Critical Infrastructure Protection and Resiliency Strategic Plan,* meant to ensure that "communities, businesses, and government are safe, secure, and prepared . . . in the event of a terrorist attack, natural disaster, or other type of significant incident" mentions terrorists seventeen times and sea level rise not at all.

"The Virginia Department of Environmental Quality does not have the expertise to study climate change issues," a spokesperson there wrote me. Bravo the City of Alexandria, then! It does have the expertise to study climate change issues, and to make a start on doing something about them.

"Sea level rise and the likely increase in hurricane intensity and associated storm surge will be among the most serious consequences of climate change," Alexandria's *Energy and Climate Change Action Plan* states. "Decreased water availability is very likely to affect the City's economy as well as its natural systems. Increases in air and water temperatures will cause heat-related stresses for people, plants, and animals.

"Quality of life will be affected by increasing heat stress, water scarcity, severe weather events, and reduced availability of insurance for at-risk properties. These impacts, some of which are already being observed, will likely have a significant effect on Alexandria's ecosystems, infrastructure, residents and economy." A sample of the plan's targets for city government: switch to renewable energy sources for a quarter of electric power consumption and reduce emissions by 20 percent by the year 2020 — by 80 percent by 2050. Make all new city buildings carbon-neutral by 2030. Charlottesville, several Hampton Roads cities, and others in Virginia have also begun to move.

THE MOST AMBITIOUS Virginia effort to date was the report of the Governor's Commission on Climate Change published in 2008. (Another commission was convened in 2014.) Of the forty-three members of the first commission, only a handful were scientists, and just one, Jagadish Shukla, specializes in climate science.

Where his resume discusses Shukla's early education, you find only the notation, "Under a tree." "It's literal," he told me when we spoke at his office. "I was born in a village in India, and there were no schools. But there was a big tree and that's where my father decided, 'We'll start a primary school.'"

Shukla's doctoral degrees are in geophysics and meteorology. He is a professor of climate dynamics at George Mason University and president of its Institute of Global Environment and Society. He has served on the UN's Intergovernmental Panel on Climate Change. His impression of the Virginia Commission is instructive, as is his prescription:

"It was an eye-opener for me, that experience. An eye-opener. For one thing, there was so much reluctance to do anything. I mean I had to get into an argument with the people who were in charge of it because they would not say anything about cutting the emissions.

"There was a large number of representatives from the business community, most of whom did not want any action. And the most — I'd say amusing or disturbing thing, I don't know what to say — is that the lawmakers will usually vote with them. Scientists, there were three or four of us. So for each of us it was really a learning experience.

"Slowly it occurred to me: I don't know if there's any limit to which the Virginia business community will go, to be able not to do anything. Everything that happened with tobacco, I just see happening with the global warming question. Finally I realized why there's no action in the U.S. Congress.

"What we are hoping is that in Virginia there will be a mitigation plan, an implementation plan. I wish that there was a follow-up group meeting: 'Okay, so what do we do?' Many other states such as Maryland are already writing or have written this plan.

"But in the final analysis, honestly, we really have to educate people so they choose leaders based on their science literacy about basic climate issues. I mean at the end of the day, big policy decisions will not be made unless our elected officials themselves understand the gravity of the situation. Because any amount of activists puffing up their chests is not going to change policy, and any amount of individual effort is not going to make a big difference.

"It's like we are conducting a very dangerous experiment on our planet. But it's an experiment, so the real consequences are going to be to our grandchildren." (Shukla's daughter and two granddaughters live in Virginia.)

His admonition that businesspeople must become better informed because they can mobilize political leaders is especially on target. Market-driven capitalist enterprise will be essential for meeting the climate challenge. Government will first have to rerig tax favors and other subsidies and incentives away from fossil fuel–based energy, though, and toward climate-friendly conservation and innovation.

That kind of government leadership has not yet materialized, and likely will not until it is demanded. Few if any of the 2008 state commission's "action plan" recommendations have been acted upon, but it is hard to be precise since no single agency has tracked their fate. They were far from revolutionary, but a useful start.

Now, a long time since, Virginia might easily join California, Connecticut, Maryland, Massachusetts, New York, Oregon, Rhode Island, and Vermont in implementing new rules that will put 3.3 million new zero-emission vehicles — plug-in electrics, hybrids, hydrogen fuel-cell cars — on the road by 2025. These states alone make up 25 percent of the national vehicle market, enough to spur reluctant automakers to adapt.

Maryland is already on its second round of bipartisan initiatives to cut greenhouse gas emissions, adapt to climate change impacts, and try to reap the economic benefits of innovation while doing so.

Despite the fanfares and high hopes that attend new legislation — thirty-seven thousand new jobs and a billion dollars of expected revenue, in Maryland's case — those benefits are not guaranteed to materialize. But they are certainly not implausible. Federal and state incentives can innovate and promote in the realm of conservation and alternative energy, just as they have for decades in favor of fossil fuels and wasteful energy practices. Texas now has more workers in the solar energy industry than it has ranchers — one result of job-creating alternative energy incentives.

Maryland's plan has other advantages, too, such as improved public health, improved land use planning and transportation, and, not least, an inspiring blow in favor of climate stability.

Its major features include:

- Twenty percent of the state's electricity must come from renewable sources such as wind and solar power by 2022. Virginia is one of a minority of states with no mandatory renewable power standard at all. Too hard? Already, one-third of Germany's electricity is from renewables.
- Reductions of electricity consumption and peak load demand by 15 percent by 2015, using numerous state- and utility-managed energy efficiency and conservation programs.
- The near-elimination of solid waste sent to landfills or incinerators for disposal — the result of reuse, recycling, composting, and reduction of the waste stream.
- A Clean Cars Program with stricter vehicle emission standards and direct regulation of CO_2 emissions. This also has the benefit of reducing unhealthy air pollutants.
- A Regional Greenhouse Gas Initiative that includes nine Northeast and Mid-Atlantic states — Virginia is not on the list — to design and implement a regional power plant emissions cap-and-trade program. Revenues from the program support energy efficiency plans.
- Public transportation initiatives to reduce per-capita vehicle miles, which have been trending upward. Transportation is a growing fraction of greenhouse gas emissions in Virginia, too.

- Managing forests on public and private lands to increase the amount of carbon captured in forest biomass, and durable wood products. This will make renewable biomass available to burn for energy production and support forest growth. There is also a goal to plant or replant 43,030 acres of forest by 2020.

In Virginia, scientists and planners need immediate, reliable funding for aggressive research on adaptation strategies. This is true for urban planners and public health agencies as well as forest, wildlife, and wetlands resource management planners. The state is not rich just now, and a carbon tax can fund these essential initiatives.

There are many other kinds of climate action that Virginia could initiate immediately:

- Enact a "carbon tax" on greenhouse gas emissions — already under serious consideration in several states.
- Protect and expand — through acquisition and tax incentives — the migration corridors and "focal areas" of ecological flow (see fig. 15) mapped out by state and Nature Conservancy scientists.
- Enact natural areas plans to aid plant and animal migrations. The plans should halt commercial logging and roading on public lands, and manage forests to anticipate a hotter climate.
- Campaign for "no-take" marine sanctuaries to protect fish stocks and the rest of Virginia's offshore estate, to add to this ecosystem's potential for survival under the threats of warming and acidification. We have almost no protection of this kind now — .02 percent of marine waters under Virginia's jurisdiction, just half of one square mile, are designated "no-take" zones. We rank far behind other states in establishing this level of protection.
- Rewrite real estate disclosure laws so that purchasers of coastal and riparian properties are given an honest picture of the threat of sea level rise, and what they will and won't be allowed to do to protect their property.

- Authorize regional planning agencies to design a phased withdrawal from shorelines as and when sea level rise advances, and establish a fund for demolition and clean-up.
- Require a "rolling easement" policy for new coastal development to enable ecosystems to migrate inland, and allow society to avoid the costs and hazards of protecting low-elevation lands from the rising sea.
- Oppose public subsidies that encourage additional coastal construction in the path of sea level rise, unless it is part of a "rolling easement" plan that protects property owners and natural areas alike.
- Establish fair and uniform state aid processes for communities affected by sea level rise or other climate change impacts. Face the mounting costs squarely, and determine how to meet them.
- Put together a comprehensive plan to protect and expand coastal natural areas. Allow only absolutely necessary shoreline hardening in areas that will be impacted by sea level rise, and empower state agencies rather than local boards to make those decisions. A companion plan would specify where we should armor the shorelines to protect real estate that is too valuable to abandon, and plan for how that protection will be financed.
- Bear down harder on controlling the pollution from cities and farms that is already killing the Chesapeake Bay, in advance of the onset of acidification from global warming.
- Lobby for a new federal inspection policy for international trade, to stop the introduction of new alien invasive plants, animals, and diseases — the kind of program that is already in place, and succeeding well, in Australia and New Zealand. Invasives, as explained in chapter 6, will be the destructive legatees of our climate-altered public and private forests and grasslands unless we stop importing them.
- Establish state-sponsored gene and seed banks for wild and increasingly rare native species, following the lead of the North Carolina Botanical Garden and the programs of many other states.

We can try to hedge our bets against the potential extinction of genotypes and species with the onset of climate change, invasive species, and other threats.

ASSUME ALL THAT GETS DONE, and something like a Maryland plan is adopted for Virginia in the bargain. Will it be enough to ward off serious climate disruption? No. But it's a beginning and a long step forward, insufficient but crucial. Our political leaders will need to be more candid with us on that score, even as they try to reckon with the art of the possible rather than with what is ideal. So if our question is not, "What is politically likely just now?" but instead, "What needs to be done to avoid the worst impacts of climate disruption?" then we return to the "climate-friendly" campaign suggested for you near the beginning of this chapter. This time, we won't pretend that you should pursue it alone.

The details of that global "work and hope" plan, the one needed to fully address climate change, are beyond my aim here. But its scale is suggested in a proposal published in the peer-reviewed *Journal of Global Responsibility*, whose arithmetic is not markedly different from others. The target is to roll back the atmospheric concentration of carbon dioxide to 350 parts per million (ppm) (recall that we're around 400 ppm now). This, the authors calculate, would likely result in a 1.8-degree rise in global average temperature. You can see in chapter 3 what the climate models project for Virginia at that level of warming.

The climate changes this will bring to the planet are judged large and negative but tolerable, and the target, feasible. This is only a speculative safe limit — other climate scientists warn that it may be too high, and some think 450 ppm would be okay. In any case, in order to progress that far, the authors calculate that we will need the equivalent of a World War II mobilization, lasting a century and beginning by 2018. They are hopeful that when disruption bites hard enough, humanity will respond successfully.

The United States, China, and Europe (later Russia, India, Japan, and Brazil) would need to initiate a crash program to reduce greenhouse gas emissions by 50 percent in the first five years and achieve "net zero" emissions by fifteen years later. It will be objected that this is an absurd, fantastic scheme. Of course. But not so demented as our current gamble that the consequences of "business as usual" can somehow be muddled through.

What we need in order to begin, though, is a change of mind-set rather than a long to-do list that won't get done. Goals and platitudes and promises will be worthless without leadership that works from a settled conviction that the climate emergency is upon us.

Here's a slogan for us, Virginia: Let's get on with it.

ORACLES

If you want to know something more about the sources of climate projections for Virginia, then Cheyenne, Wyoming, is not a bad place to start. An advance guard of 18-wheelers rolled into a business park there not long ago to unload some of the first chunks of a supercomputer called Yellowstone, a quadrillion-calculations-per-second thinker that will be used to model future climate.

Yellowstone has thirty times the power of the supercomputer that the National Center for Atmospheric Research (NCAR) and other climate teams have often depended on over the last few years. It cost about $30 million, and just like your own computer, it is likely to be obsolete and ready for replacement by a faster, cheaper unit within four years.

If you took this machine to the drag races, you could tell your friends that it sports a "Mellanox Infiniband full fat tree," among other accoutrements. At least for a brief time, this was the fifth-fastest computer in the world, according to Richard Loft, an NCAR technology director. Made by IBM, it is designed to store a hundred thousand times the data held by your laptop. It generates enough heat to require constant cooling with chilled air and water. That's because it draws about 2 megawatts of power — more, when the problems set for it are more demanding. That much electrical energy could power 1,600 average American homes.

Which explains Cheyenne. Its air is dry, which makes cooling more efficient, and its electricity is the cheapest in the United States. Just one of those commonplace ironies: some 90 percent of the new climate-modeling supercomputer's electricity comes from coal-fired power plants. As for the "full fat tree," it is a communication interconnect that boosts the speed of the conversations — to 56 gigabytes per second — among Yellowstone's 72,000 processors.

Marika Holland, chief scientist for NCAR's Community Earth System Modeling Project, thinks Yellowstone is "close to a game-changer. We've had incremental improvements in our computational resources over time, but you can't get as much interesting science done, because everybody's clamoring for those resources."

If you have an old digital camera that doesn't have as many megapixels as a new one, you say, "I want to get that camera because it takes sharper pictures, and I can store a lot more pictures on it." It's the same thing with this computer. It promises a sharper, more detailed view of the climate system and a lot more memory.

Global models often cannot "see" details that matter, such as ridges and valleys that affect rainfall and which watershed it occurs in. Complex coastlines and mountain ranges — Virginia has both — also affect climate. They show forth in more finely resolved regional models, but modeling climate on a regional scale is an especially power-hungry process.

Current climate models for North America are typically based on a grid whose units (think "pixels") are enormous. Just quartering those squares uses roughly fifty times more computing power, so improvements there have been essential. Regional models call for grid units as small as 7 miles on a side.

The climate researcher Adam Terando of the U.S. Geological Survey is refining the results of regional modeling so that decision makers can use it. How dry it will be by midcentury, say, in the watersheds that feed the Occoquan Reservoir, which provides water for hundreds of thousands of Northern Virginians. Or how hot the summers may get in the Shenandoah Valley, how quickly the Greenland ice sheet may melt and raise water levels along the Chesapeake, or where loblolly pine, a major source of timber for Virginia, can and can't survive as the climate warms.

On the other hand, there's the "garbage in/garbage out" problem. The GPS unit on the dashboard of my car is a great piece of gee-whiz computer technology, too, and its maps are meant to tell me something about the future. But on plenty of occasions, that magic box has misguided me bizarrely: to a cow pasture instead of the hotel I was looking for, to the middle of a bridge over the Potomac River instead of an office building, to the far end of the block instead of my own house. Why trust a climate computer?

Climate model reliability is not much about new, dazzling hardware. It's more about the very long strings of programming data that are fed into those chips and drives. They encode the physics of climatic processes, but also a long list of necessary assumptions.

Sigmund Freud once referred to *Homo sapiens* as "a kind of prosthetic God. When he puts on all his auxiliary organs he is truly magnificent; but those organs have not grown on to him and they still give him much trouble at times." We use telescopes as prosthetic eyes, automobiles as prosthetic legs . . . You can guess the rest of the long list. And when we want to cerebrate future probabilities for some kind of physical process like climate change, modeling might be called our oracle prosthesis.

It is far more precise and much easier just to measure climate change over time than to predict it. If we want to know what may happen in the future, though, there is no way to observe that directly. We can no longer assume it will be more or less as it has been in the past. So instead, scientists build a model of the physical processes that govern climate, and how they unfold and interact with each other over time as greenhouse gases are added in.

Aircraft designers from the Wright Brothers onward used model airplanes and wind tunnels to predict the likely impact of various changes in design. The model was not a perfect replica of the plane, just a small simulacrum that would usefully approximate the future behavior of the real thing.

As for those problems Freud alluded to, they are good to anticipate, and to plan carefully for. If we're going to invest credibility in climate models, then we need to know something about the quality-control strategies of the people who construct and use them. Several caveats about models are worth keeping in mind. You may have heard this one: "Essentially, all models are wrong, but some of them are useful."

Regional models, such as the ones in this book, should especially be approached, as I've been told by those who create or use them, with "a healthy disrespect." Which is to say that all of these colored patches on the climate maps, those mute signals of the coming heat, are science. Which is, from its Big Bang theories down to its bosons, fundamentally imprecise — a story that is always under development.

So, many gaps in our climate models remain to be filled in. "It is still really hard to gauge the future accuracy of any one model," Terando said. "There is too much uncertainty about how the climate system is going to respond to greenhouse gas emissions, particularly in a small area like Virginia. That's why we use multiple models, so we know the range of uncertainty. That's how we hedge our bets. The true answer is more likely to lie somewhere within that range of outputs."

Climate scientists have worked at narrowing the range, weaving uncertainty and a kind of disciplined humility into the climate-modeling enterprise from its early days, about fifty years ago. Sometimes the different models are considered separately, and sometimes they are combined and averaged. A reality check, then, is that the competing models in this international suite are constantly compared with each other. When they generate different results as they attempt to reproduce past climates, or project future climates, the modelers have to sweat — for years sometimes — the answer to the obvious question: Why?

Unfortunately, even the crude early models predicted the current warming of the planet with substantial accuracy, and they are unanimous that it will continue — only differing over how intensely and how rapidly. "Over several decades of development, models have consistently provided a robust and unambiguous picture of significant climate warming in response to increasing greenhouse gases," the most recent assessment by the IPCC states.

Another way to test models is to start them at some point in the past and then run them dozens of times. The resulting patterns show how accurately and how consistently the models can "hindcast" fairly recent climate conditions, ones that we already know well from the historical record — about 150 years back. They are also tested to see if they can simulate the climates of thousands or millions of years ago. Those climates have been reconstructed from tree rings, fossils, and ice cores.

A model doesn't "know" what we know about those ancient climates. The test is for the model to be able to reproduce them, based on the climate processes that are embodied in some tens of millions of lines of computer code.

Where the models agree, there is a higher level of confidence in their projections. Where they do not agree is just as important. "That allows us to quantify the uncertainty in the models because they are a representation of the real system. They cannot be as complex as the real system," Marika Holland said.

When climate scientists acknowledge uncertainties and which issues are still unresolved — and that is a commonplace in the research — it shows a discipline that takes us well away from the land of make-believe. As we've seen, climate models disagree so much for now about the coming century's rainfall in Virginia (and the rest of the Southeast) that only a few, low-confidence projections are currently on offer. That's an unapologetic part of the public scientific record, and it spurs the research.

Another quality-control strategy is nearly always employed. It's that the range of possibilities the models project are averaged over spans of time. Midcentury summer temperatures in Virginia aren't expressed as a single number in a single year's summer. Instead, the research offers a range of temperatures over several years. This is not an admission of failure. It is fundamental to how science proceeds. It blurs and complicates the story line, but it also makes it more trustworthy.

Each of the competing models uses a somewhat different set of approaches to constructing an abstract climate, a cybermachinery of mathematical atmospheres, jet streams, solar heating and cooling, and a long list of other physical processes. Atmospheric models were first worked out in the 1960s, but they did not include CO_2 in the atmosphere except as a fixed component. Adding in more CO_2 to see what would happen to the climate came next.

Ocean models — circulation, currents, temperatures, salinity — were devised at about the same time, but they weren't linked up with the atmosphere models. Land temperature models and moisture models showed up. Modeling the physics of sea ice, how it moves, grows, and melts, started in the 1970s. Gradually, these were coupled together so that they could interact.

Terrestrial system models today simulate how much carbon green plants pick up and how much they release, and how those cycles change with growth cycles and types of vegetation. Models can even include a marine ecosystem that simulates phytoplankton and their intake of carbon, which in turn clarifies the range of atmospheric CO_2 concentrations.

Some significant factors are still largely missing from the climate models, though, or are only included in rudimentary formulations. How quickly land-based polar ice-sheets melt as the planet warms, for one thing. Another question mark is the powerfully insulative greenhouse gas methane, which is now emerging in unprecedented and disturbing quantities from permafrost, from newly appearing vents in the Arctic seas, and from other new methane sources that are responding to warming.

Then, there are the cloud problems. "Well, they're very complicated," Marika Holland told me from her office in the foothills of Boulder. "I'm looking out my window right now. Clouds come in many different flavors. Thick, thin, extensive, very scattered. All of them have different impacts on how much sunlight reaches us, and how much is reflected back to space, and also on the long-wave emissions of heat."

Clouds affect how much heat remains within the Earth's system, but they're complex and they occur at small scales. They are one of the reasons behind the different projections of future climate across different models, Holland said.

A CLOSE-TO-HOME example of this modeling uncertainty involves Virginia's recent temperature record. The temperature charts in chapter 2 document, you may recall, that there was little change from the 1940s to the 1970s, and that is a lingering mystery for Virginia and other parts of the south-central and eastern United States. Much of the rest of the country, and most of the planet, was warming steadily.

The research continues on this question because it helps refine the models and makes them more accurate. Recent studies point to the causes of this "warming hole" as a possible combination of two factors, though. First, the longer-term air circulation patterns caused by the oceans. Second, the pall of pollution that still hangs over the East, much of it from coal-burning power plants, especially those in the Ohio and Tennessee river valleys.

That fine-particle smog continues to cause premature deaths each year,

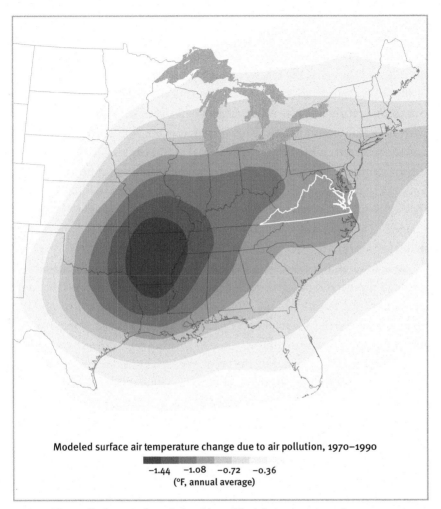

Modeled surface air temperature change due to air pollution, 1970–1990

−1.44   −1.08   −0.72   −0.36
(°F, annual average)

FIG. 24. How pollution may have helped keep Virginia temperatures low, 1970–1990

according to the Environmental Protection Agency, sickens the lungs and hearts of countless others of us, and poisons mountain soils and streams with acid rain. But it was far more severe before the Clean Air Act began to take effect in the 1970s.

Ironically, as figure 24 suggests, the coal-smoke sulfate particles may have shielded Virginia, by reflecting and scattering sunlight across the region. This is a modeled diagram by NASA of the great pollution cloud east of the

Mississippi that, at least according to one recent study, lowered temperatures from East Texas to the Atlantic region even as late as 1970–1990. By then, as we've seen, warming was clearly in evidence in Virginia, but pollution may have slowed the warming rate by about twenty years.

The air is cleaner now — a good thing — but the heat is building. The distinct temperature increase on the charts we've seen shows what may be, in part, the result of cleaner air. That question is not well resolved, however, and the research continues. A hypothetical public discussion about climate disruption illustrates how uncertainty can be embraced, if not celebrated. Let's say that at some point Earth's average temperature stabilizes, despite the climbing concentration of greenhouse gases over that period. What does that mean for the future? Could we just relax and throw all the gloomy climate talk overboard?

We could not, if we're still talking about only a few years of data within a much longer trend of warming. This revisits the "weather is not the same as climate" conversation. Still, climate scientists would be called upon to figure out where the stored energy, held in by our thickened global blanket of greenhouse gases, was going.

In this pretend conversation, the point to keep in mind is that a change in current observations — even over a few years — would not derail decades of earlier findings. That would occur for political reasons, not scientific ones. Instead, the newer data would be collected, and incorporated, over a far longer and more patient stretch of time.

For argument's sake, though, let's assume as a given the welcome possibility that the "cooler" climate models are going to prove more correct in the long run. That's some of the "hope" in the "work and hope" emissions scenario, and we will still need all the "work." Because the role that these models' lower climate sensitivity plays is limited. It will determine when, not whether, the planet will get hotter.

Lower sensitivity could buy us as much as a couple of decades to figure out how to adapt, and how to reduce emissions. But the habit of burning fossil fuels is the bigger variable in determining how much warming we'll see, and when, according to the climate scientist Katharine Hayhoe: "Here's the deal. When you get out to the end of the century, the uncertainty be-

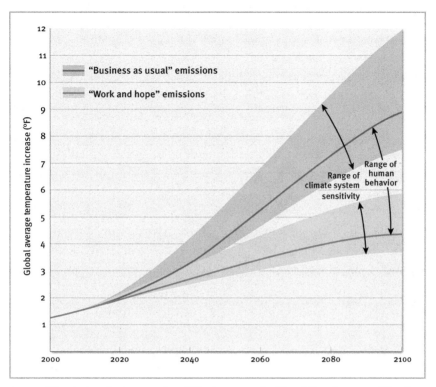

FIG. 25. The choices humans make will be far more significant than how sensitive the climate system is to greenhouse gases. (*Note:* Solid lines in each color band are the IPCC's best estimate.)

tween the higher-versus-lower emissions scenarios is much larger than the uncertainty between more-sensitive and less-sensitive models." The choice-of-model knob on the climate box is not the major factor. The major factor is us. To illustrate this point, she created figure 25 based on data from the Intergovernmental Panel on Climate Change.

Each colored path shows an assumption about how much of those greenhouse gases we'll pump into the atmosphere during the 2000s — the familiar "business as usual" and "work and hope" emissions paths.

The width of each colored path shows the range of climate sensitivity as projected by the different climate models as we move through this century. In the "business as usual" mode, it takes the "hottest," most climate-sensitive models only until about the year 2045 to add 5 degrees to the global temperature average. The cooler models don't project that much warming until twenty years later.

Note that the further we look into the future, the wider these uncertainty bands become, as you would expect. The solid lines within each colored path illustrate the IPCC's best estimate of how much global warming is likely to happen within each of these two emissions scenarios.

Our behavior makes the largest difference, no matter which models we choose to look at. As we've said, the lower "work and hope" path, in which we control emissions tightly, is increasingly hard to reach, like a bank account of options that we keep spending down. "The opportunity to get out of trouble is dwindling," Hayhoe said. "We are entering into unknown territory where humans control the thermostat of the planet. We cannot predict what humans are going to do."

THE VARIOUS VIRGINIA projections you've seen in this book rest on an important assumption: that what we know about the climate systems of past centuries will operate much the same way in the future. That the general rules for how polar ice, ocean currents, or the jet stream behave will still govern what happens next — that's called "stationarity."

Climate disruption may invalidate that assumption. Feedback mechanisms that have not yet been accounted for, such as the methane now escaping from thawing permafrost, or CO_2 generated from an acceleration in the number of forest fires around the planet, may mean that the rules will turn unruly. The models do not — really, they cannot — always take such wild cards into account, because they are not yet known.

The climate scientist Gavin Schmidt has appropriated a phrase made familiar by former defense secretary Donald Rumsfeld to describe the stationarity problem: climate models cannot contemplate "unknown unknowns." That is, many of our well-learned expectations about how global climate works could be dumped overboard by a new, destabilized climate.

In the words of the most recent federal climate assessment: "Some changes may occur in a relatively predictable way, while others involve unexpected break-points or thresholds beyond which there are irreversible changes or changes of higher magnitudes than expected based on previous experience. These 'tipping points' are hard to predict." Which is to say: modeling projections look smooth, but reality can be abrupt.

Not often heard in the talk about models, hardware, and data is a different concern, outside the ambit of climatology but distinctly on the minds of climatologists: how people will cope with what's coming. "For me, it is very concerning," Holland said. "I'm a mom. I have two kids, and it does worry me, the planet we might leave for our children and their children. I guess I'm also an optimist, though. I think that we as a society can make decisions and make the right choices, so that's where I try to find solace.

"So how do I deal with it? Well, I think it's kind of day-to-day. The work I'm doing is very interesting for me, and I hope that it can be used to make sensible decisions. But the truth of the matter is it takes more than a handful of individuals doing that. It needs to really be everyone. I don't know how to make that transition, so my perspective on it is I do the best science I can, and provide the best information I can. At the end of the day, I hope that makes a difference."

NCAR's Richard Loft said he is deeply concerned about projections that raise the strong possibility of unprecedented drought through large regions of temperate land masses, such as the central and western United States — far worse than the drought that has already caused water rationing and the liquidation of cattle herds in California and Texas. "Would you rather go to a doctor who tells you some happy horseshit, or what's really going to happen?" Loft asks. "Not everybody has the same answer to that, I guess. I'd like to know.

"Dinosaurs didn't have any technology to know something was coming, so they just woke up one day and found out that the planet was destroyed. If we can know about it thirty years before it happens, and do some things to cope with it, we can still mitigate the impact on human beings."

Unavoidably, though, more knowledge about climate change leads us directly to difficult choices — trade-offs, sacrifices, opportunities. Who among us, for instance, should sacrifice more in order to cut back on greenhouse gas emissions? How long do we wait before we begin to act in earnest? As we make those decisions, we learn as much about ourselves and our society as we do about climate change.

Karen Akerlof, part of George Mason University's Center for Climate Change Communication in Fairfax, reminded me once that consulting an oracle is a deep — and ancient — need: "In other eras, we asked the gods these questions instead of main-frame computers. But in mythology, the answers from the gods typically did not solve the protagonist's problems.

"Modeling has the same issue in relation to policy today. Striving for greater model accuracy — and understanding of the future — may never get us what we think we 'need' to make policy decisions," she said. In other words, we may always question whether the science oracle gives information that is certain enough to act upon. In the next chapter, we'll give that idea a longer look: Why can't science decide what we should do?

A HIERARCHY OF CREDIBILITY

Hundreds of books and websites offer proposals on every scale, from planetwide to personal, for reducing our carbon footprint in order to ease climate disruption. To make that torrent of conversation useful, though, we have a different challenge. How do we decide whom to listen to in the future, about either the fixes or about the climate change itself? Some of those "solutions" are bogus, and others threaten to do even more environmental harm. We need a rough guide to help us sort out what is, and is not, believable. Really, now, why believe what you are reading here?

One way for those conversations to go off the rails is the confusion of science issues with other kinds that are based instead on our values. That problem has been usefully explored by a University of Colorado political scientist, Roger Pielke Jr., who comes with an interesting pedigree.

For one thing, he is frequently at odds with mainstream climate scientists. His father, Roger Pielke Sr., is himself a climate scientist at the University of Colorado. He, too, often contends against some of the mainstream scientific view. Much of that mainstream research emanates from the National Center for Atmospheric Research, which is federally funded but has close ties to the University of Colorado.

So Pielke Jr. in his book *The Honest Broker* explains that questions about climate science are answered by research based on the scientific method. Many other questions cannot be.

This is a science question: How much will the level of the sea rise along Virginia's coastline in the coming century? Here's a quite different sort of question: What should Virginia do about sea level rise? Pielke calls those "value" or "political" questions. They have to do with our beliefs and priorities. They can be raised as a result of scientific findings, but they can't be answered with the tools that science uses.

Scientists can help as "honest brokers" at times, providing data about the impact of different courses of action: What is likely to happen to wetlands if we build seawalls from Point A to Point B? What is likely to happen if we do not? But science cannot choose among those alternatives for us. Weighing pros and cons of that kind is the job of our political system.

Pielke writes that inevitably, some scientists decide to advocate certain policy choices about, for example, climate change. If they do, it may affect how their research is viewed by the public. For value choices, Pielke argues, scientists bring no more nor less to the table than do the other thoughtful, well-informed, experienced people we may want to heed.

Whatever role a scientist adopts, the techniques of science research cannot substitute for the political process. When science questions are ensnarled with value questions in a political discussion that does not carefully distinguish between the two, Pielke writes, we are likely to confuse ourselves in a damaging way.

If we're trying to decide whom to listen to about the science part of the climate change discussion rather than the policy-making part, then we would do best to pay attention to scientists rather than movie stars, talk-show hosts, or political representatives — unless they themselves are relying on credible science. Figuring that out is not as difficult as it may sound. A strategy, a hierarchy of credibility, is what this chapter proposes.

Anyone can offer an opinion or pass along scientific conclusions, as I am in this book, but not everyone is qualified to "do science." In sum, here's how this works: researchers submit their results to referees — their peers in whatever discipline they pursue — at science journals. The research is carefully reviewed, and often rejected or sent back to its authors for revision. If it's judged to be worthy science, it is published.

Then it's subjected to the criticism or provisional acceptance of the wider scientific community. As time passes, this peer-review process tests hypotheses and data continuously. Some hold up. Some are superseded by new findings that are more consistent and reliable. Science research is an ongoing conversation, not a set of static conclusions.

When the time arrives to discuss policy — the value questions — the arena broadens to include Al Gore or Rush Limbaugh. First judge whether they adhere to the science you generally trust, and then you can weigh their policy arguments. What does the Virginia state delegate or the congressional representative you voted for say about climate science, and where does that opinion really come from?

As you already know, however, processing the climate issue is not quite as tidy as all that. Let's move the discussion here: How do we decide who has more credibility when the climate scientists themselves disagree about the science?

Despite a strong consensus among climate scientists that goes back at least thirty years, there are many reasons for the divide in public opinion about climate change. The "controversy" is political, not scientific. A handful of climate scientists have always questioned the mainstream, but usually in statements that are not part of published, peer-reviewed research. That's a valid contribution, but if it's political, it's not science. And if it is science, then we have to ask why it hasn't been published in a reputable science journal. If science research does not undergo this form of quality control, it rarely alters an overwhelming scientific consensus, and for good reason.

Every major U.S. scientific group that addresses climate questions has agreed, formally, on the record, that climate change is under way, that it is human-caused, and that it will have very far-reaching impacts: the National Academy of Sciences, the American Meteorological Society, the American Geophysical Union, and the American Association for the Advancement of Science among them.

Take these studies into account:

Of the 928 climate research papers published in refereed scientific journals between 1993 and 2003, 75 percent either explicitly or implicitly accepted the consensus view on global warming, 25 percent took no position, and none disagreed with it.

In another study, 3,146 geochemists, geophysicists, oceanographers, geologists, hydrologists, and paleontologists were asked whether they think human activity is a significant contributing factor in changing global temperatures, and 82 percent said yes. More than 90 percent had Ph.D.s. The survey included participants with well-documented dissenting opinions on global warming theory. The more climate science research they did, the more likely they were to answer the question in the affirmative.

Of 4,013 climate research papers expressing a position on global warming, 97.1 percent endorsed the consensus position that humans are causing it. "Our analysis indicates that the number of papers rejecting the consensus on anthropogenic global warming is a vanishingly small proportion of the published research," the study's authors concluded.

Then 1,372 climate researchers and their published research were analyzed in yet another study. Ninety-seven percent of the ones most actively publishing in the field support the views of the Intergovernmental Panel on Climate Change on global warming. The expertise and scientific prominence of the researchers who are not convinced of climate change were substantially less than those of the mainstream researchers.

These studies all appeared in peer-reviewed journals and were sometimes buffeted, before and after publication, by responsible criticism. That's the quality-control process, imperfect but a tall step up from the unsifted claims and snarky rants of much of the TV commentariat and the blogosphere.

AMONG CLIMATE SCIENTISTS, there is a spectrum of opinion that is sometimes hidden behind all the polarized policy rhetoric and labels like "climate denier," "warmist," "skeptic," or "apocalyst." There's the tiny minority that I will call "contrarian" for convenience, and the group that sees climate change as a profoundly destabilizing challenge: the "mainstream." Even the contrarians, these days, are only rarely if ever heard to argue that there is no global warming, or that it is not the result of greenhouse gas emissions.

In Virginia, no conversation about this issue would be complete without the voice of climate scientist Patrick J. Michaels, the state climatologist for twenty-seven years, a former faculty member of the Department of Environmental Sciences at the University of Virginia, and a longtime campaigner in the international "climate wars." He is of interest in this conversation not because of his history, though. Instead his example and arguments are instructive for the future, as we nonscientists work out some ground rules about how to judge what we're being told about climate science, or climate solutions.

Michaels affirms that global warming is under way, but he questions much of mainstream science. He looks at the data I've recited above about the consensus among scientists and sees an involuted group of closed minds — climatologists who reflexively exclude dissenting research findings from the scientific journals they control.

Michaels's research papers and outspoken books, op-eds, interviews, speeches, and congressional testimony on behalf of the "contrarian" view of climate change have commanded plenty of news coverage over nearly three decades, to the discomfiture of his detractors. They point out, correctly, that in this way the news media leave the public with the misimpression that there is a controversy about the science of global warming, when in fact none really exists except at the furthest margins.

It's not a polite conversation out there at times. Michaels has been the focus of withering criticism from both scientists and nonscientists as a "denier" and a "disinformer," a "fraud," and a "serial deleter of inconvenient data," and he and his allies have returned the disfavor, with interest. His online *World Climate Report* sums up his outlook: "Climate change is a largely overblown issue.... [T]he best expectation is modest change over the next 100 years."

Michaels believes that government-sponsored research promotes alarmism, in order to groom careers and attract more funding. He is employed now at the Cato Institute, a libertarian think tank based in Washington. His published work is usually about policy choices, though he has also contributed research to peer-reviewed journals.

I posed our credibility question during a conversation with Michaels: How can citizens with just average science literacy make sense of competing scientific claims about climate change? He suggested a couple of opposing blogs they could monitor, one mainstream, one contrarian. From them, he said, readers could weigh the arguments and synthesize their own views.

I'm familiar with both blogs (realclimate.org and wattsupwiththat.com), but this was a surprising suggestion in some respects. I find each of them opaque. The first was far more focused on science — it is written by climate scientists — but the other is focused more often on invective — it is written by a former television weatherman who is listed as one of several coauthors on only two published research papers. The discussions on both sites are recursive, recondite, far out in the weeds.

The suggestion was immediately understandable in another way, however, because it promotes what Michaels's critics refer to as "false equivalency." A generation of news reporters trying to make sense of climate issues has often fallen for this: if two points of view are on offer, then be safe and treat them as having equal value.

If you're a political reporter covering two candidates and trying to remain open-minded, this can make some kind of strategic sense. It can seriously distort the truth when two obviously unequal candidates are running, though. It's as if a four-year-old and a center for the Virginia Tech Hokies are standing side by side, and, fearful of charges of bias by their fans, you describe them as being of equal height, weight, and talent.

In matters of science, false equivalency in the news has meant elevating the judgments of a tiny number of climatologists to the level of an encompassing number of their colleagues, in the "two sides to every story" mode. It leads to the false conclusion that there is a high level of scientific uncertainty about whether climate change is real and a grave threat. That has resulted over the years in severe public confusion and paralysis in policy. It has meant that without quite knowing it, we've often wound up talking about science in exclusively political terms.

This alone does not allow you and me to judge the quality of "contrarian" science like that of Patrick Michaels. It does suggest that what most of the public — including a recent president of the United States and a governor of Virginia — regarded as "uncertain," "unsettled," or "controversial" climate science was none of those things. Scientific conclusions are always to some degree provisional, subject to revision — that's the essential nature of the scientific method. But it is important to know where, on the spectrum of scientific research about climate disruption, the heaviest representation is.

We always sympathize with the outsiders and the underdogs, though, especially when they prevail and become heroes. You have to remember Galileo, who faced torture and was finally silenced because he upheld the truth and opposed orthodoxy. You have to think about Einstein, whose theory of relativity was mocked at first. And I like to ponder the case of Daniel Schechtman, an Israeli chemical physicist who won the Nobel Prize a few years ago for discoveries he had made thirty years earlier. His work on the structure of crystals, too far outside the realm of perceived reality, was for a long time uniformly rejected by colleagues: "Go away, Danny," one senior colleague told him. An earlier Nobel laureate was staunchly dismissive.

This is sometimes characterized as a juxtaposition of "mavericks versus mules," workaday science versus the inspirations of contrarian geniuses. But it's not at all clear that the exceptions are more than just that. As we'll see, acknowledging those exceptions doesn't explain away some of the other problems in the arguments—and the portfolios—of many of the climate-science "contrarians."

Who is to say that Patrick Michaels or some other contrarian won't be proven right about climate disruption in the end? Let's hope and pray, if you're a prayerful person, that that will be the case—but that is the wrong question to ask given where we find ourselves now.

The useful question is this one: where are our best odds, today and tomorrow, of finding the truest approximation of what's going to happen—especially when we can't afford to just wait and see? Often, the mules of research are plugging along, doing the productive work that pays off in spectacularly successful science we now enjoy the benefits of, in thousands of ways. Often, I proposed in a conversation with Michaels, the mavericks are completely wrong. Sometimes the outsiders are outside for good reason. How do we know which case we're dealing with? What's the quality-control process for contrarians?

"I get that question from Cato administrators all the time," he said. "How do you know what's good and what's kind of way off? My answer to that is not a very satisfactory answer. If you have a pretty sophisticated education and critical insight, particularly if you are trained in science, you can kind of tell the things that smell funny as opposed to those that may be extremely innovative." For citizens, though, that sounds like it means "Listen to Patrick Michaels's intuitions." As for most climate scientists, they are content to rely on the peer-review system instead—despite its frailties, which we'll discuss momentarily.

THERE'S ANOTHER REASON to question many of the climate science mavericks: conflict of interest, also known as divided loyalty, or "serving two masters." Scientists who take money from interested parties like oil, coal, or electric utility companies produce research data that may be viewed as mere partisan thrusts, especially when they are not submitted to peer review. It helps, at least, if these affiliations — which are certainly relevant as we try to judge both their research and their policy proposals — are made clear. Often the connections have been carefully hidden, instead.

Michaels has some intriguing suggestions about this problem, which we can explain further by way of example. Physicians are prohibited by federal law from referring Medicare patients to a medical facility — a heart-scan clinic, say, or a laboratory services corporation — in which the physician has a financial interest. That law acknowledges the temptation for doctors who might overprescribe heart scans or lab tests in order to reap more income.

Reputable news media — the *Roanoke Times* is the source of this example — have codes of ethics for journalists — both reporters and opinion writers — that also anticipate the issue:

> Staff members may not enter into a business relationship with news sources, use inside knowledge about businesses for personal gain or give anyone outside the news department knowledge of any proposed or pending story that could affect the price of securities or contracts.
>
> Financial investments, loans or other outside business activities that could conflict with the newspaper's ability to fairly and impartially report the news, or that would create the impression of such a conflict, must be avoided.

Note that even the *appearance* of a conflict is an immediate problem for credibility. This is common sense, isn't it? If you hire an architect who specifies materials for your new home, you don't want her to accept big holiday gifts from plumbing suppliers, or free vacation cruises from window manufacturers, even if she swears to you that despite how things look, she is absolutely trustworthy.

You don't want a Supreme Court justice to go duck hunting with one of the parties in a case he's hearing, or Virginia legislators to take all-expenses-paid-by-the-mining-company trips to France, when they're deciding whether to lift a ban on uranium mining. Those situations, too, have occurred. And the principals all figured, no doubt, that their judgment couldn't be bought, so everything was fine. Etcetera.

In science, as in medicine, journalism, architecture, and public service, the best cure for conflict of interest is an outright, vigilantly enforced ban. Sometimes, however — as in the case of Virginia legislators who take gifts from lobbyists — only disclosure is legally required (and not a lot of it, at that), and the only remedy relied upon is the indignation of a sporadically informed public.

In the science-and-policy arena, there is no such transparency at times. The science historians Naomi Oreskes and Erik M. Conway have confirmed in nearly endless detail the success that the tobacco industry, the fossil fuel industry, and others have enjoyed as "merchants of doubt," secretly recruiting and funding scientist-spokespersons to make a dubious but good-looking case for scientific "uncertainty" on issues such as the dangers of smoking and of secondhand smoke, or of acid rain, the ozone hole, and global warming. Sometimes the same scientists preached "uncertainty" in several such controversies, notably cancer and smoking, and global warming.

Similarly, many of the most vocal of contrarian climate scientists and pundits have accepted, sometimes secretly, tens of millions of dollars from fossil fuel industries and their political allies to promote the idea that there's a high level of scientific uncertainty about climate disruption. In recent years, the advocacy organizations, think tanks, and trade associations have concealed the sources of their funding even more actively, according to research by the sociologist Robert J. Brulle of Drexel University in the journal *Climatic Change.*

Patrick Michaels has been more forthcoming. When he was the Virginia state climatologist, he began publishing the highly opinionated *World Climate Report,* which put readers on notice that it was funded by the Western Fuels Association. In an affidavit filed in order to withdraw from a court case as an expert witness in 2007, however, Michaels declined to reveal the names of clients of his consultancy, New Hope Environmental Services, citing their need for confidentiality and the hazard to his income.

The names of two clients whose support totaled $150,000 did appear in the affidavit: Tri-State Generation & Transmission Association and the Intermountain Rural Electric Association, both electric utilities groups. Michaels confirmed that he has also been paid over the years by the Edison Electric Institute, the German Coal Mining Association, and the Cyprus Minerals Company.

Here is an exchange between Michaels and the journalist Fareed Zakaria on CNN:

> FZ: Let me ask you what people wonder about advocates like you . . .
>
> PM: I'm advocating for efficiency.
>
> FZ: Right, but people also say you're advocating also for the current petroleum-based industry to stand pat, to stay as it is, and that a lot of your research is funded by these industries.
>
> PM: No, no. First of all, what I'm saying is . . .
>
> FZ: Is your research funded by these industries?
>
> PM: Not largely. The fact of the matter is . . .
>
> FZ: Can I ask you what percentage of your work is funded by the petroleum industry?
>
> PM: I don't know. Forty percent, I don't know.

Do these conflicts of interest invalidate or compromise Michaels's published research or his voluminous public commentary on climate change? For some readers, they do indeed. That's why, in other realms, the rules about conflict of interest and disclosure are so stringent. Oddly, journalists whose jobs would be at risk if they themselves had such conflicts have often failed to inquire about or publish the sources of funding for climate change dissenters like Michaels. It cannot be demonstrated that he would have believed or behaved any differently if he had not taken the money. But at a minimum, consistent and thorough disclosure of conflicts seems requisite for us information consumers.

Contrarians argue that mainstream climate science, mostly funded by government and by universities, has its own conflicts of interest. If I'm a scientist whose money comes from there, they assert, then the more alarming my research is, the more Congress or some government agency comes up with funding, and the more my career zooms.

"So we're making a hypothesis," Michaels says. "It's the one that was leveled against me: well, you're just biased because you're supported by blah blah blah . . . What we're saying is that all of them set up conflicts of interest, not just industry but the government, and I have yet to see anyone demonstrate that the government is more moral than industry."

This may sound plausible at first. Examined more closely, however, the logic frays. It means that any employed scientist has a conflict of interest just by virtue of taking a salary. But some employers in climate research — a university or government agency — have a better claim to serving the public interest than a coal company or an electric utility, don't they? Their scientists are also afforded at least some protection against outside pressure, in the form of academic tenure and civil service rules. So not all potential conflicts of interest exert equal temptation. It's not a question of which is more moral, industry or government. It is, rather, which one of them has the biggest axe to grind, the biggest conflict between the interests they serve.

We all occupy a long list of roles, and we have a mostly intuitive sense of which ones set up a conflict that matters. If I were a climate scientist, a female Catholic mother of three in Southside, and a longtime member of the Garden Club of Virginia, could any of that significantly affect my research? As a science journalist, here's my own disclosure for you to mull: I am an occasional contributor to environmental groups and Democratic candidates. I have never accepted a dime from any of them.

Identifying with those lonely maverick heroes in the history of science, as contrarians sometimes do, also tangles up their talking points on conflicts of interest. Galileo wasn't subsidized by the telescope industry to promote a sun-centered view of the planetary system. Pasteur's research was paid for by the government. And I suppose you could complain — but not very convincingly — that Einstein was just "grooming his career" by touting alarmism via relativity theory.

Indeed, the "everyone's conflicted" argument, taken at face value, makes the claim that any science can be reduced to cheap opportunism unless it's exalted by hindsight. We have better ways to sort the credible science from the questionable. We can see more clearly than to believe that a worldwide conspiracy of self-interest is afoot among climate scientists, or perhaps an epidemic of hysteria.

Rear Admiral David Titley, a Ph.D. oceanographer and the oceanographer and navigator of the U.S. Navy, has heard that one. Titley commanded the Fleet Numerical Meteorological and Oceanographic Center in Monterey, California; was the first commanding officer of the Naval Oceanography Operations Command; and is director of the Navy's Task Force Climate Change. Among other ranks and titles. In a speech at Old Dominion University in Norfolk, he spoke of the importance of preparation for climate disruption, and for coming to terms with uncertainty by quantifying risk, which he called "counting cards in nature's casino."

"How do you deal with the skeptic who says, I don't buy it?" Titley was asked. "What I do is kind of try to just walk through the physics," he replied. "We know the climate's changing. There's just too much data. . . . One of the things you learn as a navigator — and if you don't learn it, you're probably not going to be one for very long — is that you never trust any one individual data point. I look at climate change like that.

"If it was just the Arctic, and all the glaciers were coming down but there was no temperature change and no acidification change in the ocean, well, okay, we'd probably have to look somewhere else. But when you look at all the data, it all lines up with a consistent picture. You can actually model it. You take out the CO_2, you don't get the right answer. You put in the CO_2, you get the right answer.

"What I've found is that with people who really, really believe it's a vast conspiracy theory, there's not much you can say," he has concluded. How do you get thousands of academics around the world to conspire, especially since if they could really disprove climate disruption, they would be world heroes? "Do they all sign an oath or something, that they're all going to say this?" he asked.

The contrary, "dissenting" scientific arguments that dismiss or downplay global warming and its impacts have not been summarily ignored. Patrick Michaels's research, for example, has been published in science journals — more than seventy papers, he said. Those arguments have been weighed over decades now, following the quality-control steps that are part of good science.

Public discussions by scientists and the rest of us outside of the peer-review process are also valuable, in fact essential, as long as money and deceptive practices don't dominate. We nonscientists ultimately have to decide what to do about the issues that science raises. Predictably, some of that conversation is bankrolled by corporations with a financial stake in public perceptions, though, and by people with strong political convictions. They are really in the business of persuasion, however, not science.

Much has been published to support contrarian viewpoints, often by the same handful of scientists who have ties to the fossil fuel industry or its political friends. The work just doesn't happen to be ratified by scientific evidence. Those are politically oriented discussions, made to look like science.

Republican Senator Olympia Snowe (Maine) and Democratic Senator John Rockefeller IV (West Virginia) once explained such influences in an open letter to one of many longtime corporate funders of this kind of material: "The Internet has provided Exxon-Mobil the means to wreak its havoc on U.S. credibility, while avoiding the rigors of refereed journals. While deniers can easily post something calling into question the scientific consensus on climate change, not a single refereed article in more than a decade has sought to refute it."

Peer review is imperfect — it is, after all, performed by men and women. As researchers are quick to say, they're not always right and not always certain, in any case. And they can be partisan. Here's what an infectious disease specialist told me about research in that field on climate change impacts: "It's so polarized right now that you get strident people on each side who, depending on what you say, will recommend that your paper never see the light of day if they don't like what you are saying. It has gotten very politicized." Hardly reassuring. But the task for us citizens isn't to look for a guarantee of truth. It's to calculate credibility: where the best chances for getting at some kind of provisional truth really are. That's what the scientists themselves do.

The science process just improves their odds — and ours — that when we take action, we'll be responding to our best estimation of the clearest picture of reality. We may prefer the fake-science certainties offered in political screeds, corporate PR gaming, talk-radio ridicule, or religion. We may wish, powerfully, to ignore the problem, or argue it away. We have already done that, to our considerable peril. Or we can ask some questions about the news coverage we read or hear, or the commentators we attend to. If their coverage doesn't answer these questions, it is at best defective and at worst, untrustworthy.

> Caveat lector (let the reader beware):
> Is it based on science research, or is it a statement about values and policies?
> Was the science published in a reputable, peer-reviewed science journal, and how much value have those peers assigned to it?
> Do some experts in the field — that is, those whose work has also been published in peer-reviewed journals — find fault with the new research, or with the science that underlies a policy proposition?
> Are the research scientists whose work is under discussion or part of a policy commentary transparent about their funding sources?
> Do those funding sources pose a conflict of interest for the researcher or the commentator?

If the independent and peer-reviewed research brightens the outlook on global warming or its impacts much over time, I'll be the first to celebrate, with the most profound gratitude. But it hasn't so far.

THE LATE Elisabeth Kübler-Ross, an eminent Swiss psychiatrist, would have been likely to sense our problem with climate research. She founded a hospice that cared for the terminally ill in exquisite Highland County, Virginia. She became well known in the wider world for her behavioral model of the "stages of grief."

She observed thousands of cases of people who were dying, or whose lives were going to be traumatically changed. She found that often, the last step in the grieving process is acceptance of reality, the willingness to move on. It can be preceded by, among other responses, depression, bargaining, and anger.

A steady focus on climate disruption can qualify as traumatic. It is sometimes alleged that climate activists exaggerate in order to grab public attention, but I think a different psychological strategy is far more common among them: to downplay the threat in order to avoid inducing paralysis and fear in their audiences. If you're like me, the first kind of coping that sometimes comes to mind, even now, is: "Global warming is preposterous. Can't happen." Kübler-Ross nailed it. The first stage in the grieving process, she said, is denial.

Once we accept this new state of things as real, however, we can begin to act on whatever we can still control, and to make suitable choices for dealing with the inevitable. Much conversation about global warming until now has been about how we can prevent it. Now the discussion is also about how we keep this from getting worse than it is already fated to become, and how we should adapt in order to minimize suffering.

Years ago, I gave an evening talk on mountain ecosystems to a few folks and a lot of empty chairs at a camping equipment store. There must have been a basketball game on TV. Or something. A friend of mine was perched in the back row, head cocked and arms folded, while I wrapped up with some comments on the likely effects of climate disruption.

"Oh, boo-hoo!" he said, and rather loudly. I guess he had heard something overpoignant in my delivery. His point was sharp, but funny and accurate. We have only so much room for lament when action is called for.

There is plenty of scope for that. Virginians, and the rest of us Americans, can and should push for it but we don't need to wait for a comprehensive federal response, or for individual citizens to voluntarily undertake sacrifices, though that will be welcome. We don't even need to get out in front of other states — they've already shown the way. As the ecologist David Orr has said, "Hope is a verb, with its sleeves rolled up."

NOTES

1. A Climate Conversation

1 *About a dozen trains a day:* Sullivan, CSX Corporation correspondence.

1 *Saltville fossils:* McDonald interview, correspondence.

1 *"Each one weighs":* Sullivan, CSX Corporation correspondence.

2 *Some 300 million years of subterranean pressure:* Virginia Department of Mines, Minerals and Energy, "Coal in Virginia Fact Sheet."

2 *Pound of coal:* www.eia.gov/tools/faqs/faq.cfm?id=667&t=2.

2 *Half the electrical energy:* U.S. Energy Information Administration, "Virginia Profile."

2 *The one nearest . . . 40 tons of coal an hour . . . 950,000 tons of atmospheric CO_2 a year:* Cogentrix of Richmond/Spruance Genco plant; the figures are from Jonathan Cogan, U.S. Energy Information Administration, to author, e-mail.

2 *A hundred thousand years or so:* Archer, *The Long Thaw,* 11; Bernstein et al., *Assessment Report 4,* 47.

2 *Holdren has explained:* Holdren, "Meeting the Climate-Change Challenge," 5.

3 *Sources of energy:* U.S. Government Accountability Office, *Report to the Chairman,* 21; International Energy Agency, "Global Carbon-Dioxide Emissions Increase"; U.S. Energy Information Administration, "Virginia Profile," Feb. 2016, http://www.eia.gov/State/print.cfm?sid=VA#63.

3 *Coal even faster than the others:* International Energy Agency, *Medium Term Coal Market Report 2013, Executive Summary,* 11.

3 *Show no sign of diminishing:* U.S. Department of Energy, "International Energy Outlook," 2.

3 *"Global warming is still seen":* Leiserowitz et al., *Climate Change in the American Mind,* 9.

4   *Virginia ranks seventeenth:* U.S. Department of Energy, "State-Level Energy-Related CO_2 Emissions," 10.

5   *Americans easily among the planet's biggest greenhouse-gas polluters:* United Nations Department of Economic and Social Affairs, "Greenhouse Gas Emissions per Capita." http://unstats.un.org/unsd/environment/air_greenhouse_emissions.htm.

5   *Twice the CO_2 emissions of Europeans:* Hill, "Windmills, Tides, and Solar Besides," 10103.

5   *Special interests and their money:* Appalachian Voices, Virginia Chapter of the Sierra Club, and Chesapeake Climate Action Network, *Dirty Money, Dirty Power — How Virginia's Energy Policy Serves the Interests of Top Campaign Contributors.*

5   *After all, 40 percent of the increase:* National Oceanic and Atmospheric Administration (hereafter NOAA), "Atmospheric CO_2 at Mauna Loa."

6   *Multiyear drought has reduced the corn crop:* U.S. Department of Agriculture, "National Agricultural Statistics Service Crop Production 2012 Summary," 1, 79.

6   *Those models don't capture much topography driving local climate:* Meehl interview.

2. Virginia's Climate Now

8   *Jefferson's comments on changing weather:* Jefferson, *Notes on the State of Virginia.*

9   *Stable since about eight thousand years ago:* Gill interview.

9   *Its 462-by-201-mile expanse:* www.netstate.com/states/alma/va_alma.htm. Sources disagree, however.

9   *Annual precipitation averages:* Stenger correspondence.

10 *"A climate condition typical of one region":* Hayden and Michaels, "Virginia's Climate."

12 *Different states, different ways:* State Climate Office of North Carolina, "Staff."

12 *Dishpan experiment:* Riehl and Fultz, "Jet Stream and Long Waves," 215–31.

13   *Fourth-warmest on record:* NOAA, "National Overview–February 2012," www.ncdc.noaa.gov/sotc/national/2012/2.

13   *The 328th consecutive month:* Vincent correspondence.

17–20   *Hinge-fit analysis:* Livezey et al., "Estimation and Extrapolation of Climate Normals and Climatic Trends," 1759–76.

3. Back Porch, Forward View

21   *What do models get right:* Bernstein et al., *Assessment Report 4,* Synthesis Report, 2007, 40.

22   *"Work and hope" and "business as usual" scenarios:* The "business as usual" emissions scenario used in this book is the IPCC's A1FI scenario. The "work and hope" scenario is B1, in which the major defining factor is that emissions are aggressively controlled.

26   *"Increasingly bleak":* Hayhoe and Farley, *A Climate for Change,* 117.

26   *"Having a relationship with the God of the Universe":* www.pbs.org/wgbh /nova/secretlife/scientists/katharine-hayhoe/.

26   *For figures 11, 12, and 13, these eleven models are the bases:* CCSM3, PCM, CGCM3 T47, CGCM3 T63, CNRM-CM3, ECHAM5, ECHO-G, GFDL CM2.0, GFDL CM2.1, HadCM3, HadGEM1.

30   *Midrange models project that average global temperatures are likely to climb:* These year-estimates are derived from figure 25 of this volume; and from Betts et al., "When Could Global Warming Reach 4°C?," figures 2, 7, 9.

31   *Exceeds all deaths from hurricanes:* Centers for Disease Control and Prevention (hereafter CDC), *Extreme Heat.*

33   *The data on rainfall already seem to show a trend:* Mitchell et al., *Recurrent Flooding Study,* 88.

34   *"Adverse impacts to crops":* NOAA, *Draft National Climate Assessment,* 9.

34   *Occasional bouts of extreme cold:* Bachelet interview.

35   *Atmosphere has not included this much CO_2:* Dowsett interview.

35   *Hundreds of times faster:* Butler interview.

35 *Since 56 million years ago:* Cui et al., "Slow Release of Fossil Carbon," 481–85. This research article states that current rates of CO_2 increase were likely much higher than they were 56 million years ago, but that the findings are indeterminate.

35 *First-known primate:* Ni et al., "Oldest Known Primate."

4. The Clock of the Bay

36 *Acidity increasing at a faster rate:* Hönisch et al., "Geological Record of Ocean Acidification," 1062.

36 *Report requested by Congress:* National Research Council, *A National Strategy,* summary at "Key Messages."

36 *Values expected by 2100:* NOAA, www.pmel.noaa.gov/co2/file/Percent +change+in+acidity.

37 *Oysters as large as 13 inches:* Wharton, *The Bounty of the Chesapeake.*

38 *In a week's time:* Webster and Meritt, "The Future of Oysters."

38 *Reduced by more than 98 percent:* "The catch peaked in Maryland at 615,000 tons in 1884; it has declined to about 12,000 tons in 1992" (Rothschild et al., "Decline of the Chesapeake Bay Oyster Population," 30).

39 *Sharp reduction of the grasses' coatings:* Arnold et al., "Ocean Acidification."

40 *Values of 7.8–7.9 are expected:* Caldeira and Wickett, "Ocean Model Predictions"; K. Caldeira and M. E. Wickett, "Ocean Model Predictions of Chemistry Changes from Carbon Dioxide Emissions to the Atmosphere and Ocean," *Journal of Geophysical Research* 110 (2005): 1–12.

5. Virginia's Ocean Estate

42 *Mysterious and little-known animals:* Roberts et al., *Cold-Water Corals,* 6–14.

43 *Some 60 miles off the coast:* Weisskohl correspondence.

43 *Big enough to swallow the city of Richmond:* Chaytor interview; http:// quickfacts.census.gov/qfd/states/51/51760.html.

43 *"No take" sanctuaries:* All comparison data are from the Marine Conservation Institute, and Mission Blue, *Seastates: How Well Does Your State Protect Your Coastal Waters?* 2013, 5.

45 *Waning of the last ice age:* Butler interview.

45   *Last 300 million years:* Hönisch et al., "Geological Record of Ocean Acidification," 1062.

45   *"Unknown territory":* Ibid.

46   *Sea surface temperatures:* NOAA, Northeast Fisheries Science Center, *Science Spotlight.*

46   *Fish are responding:* Nye et al., "Changing Spatial Distribution," 111–29.

47   *Decline of plankton:* Friedland et al., "Thermal Habitat Constraints," 1.

49   *Virginia is the third-largest marine producer, and state data:* Virginia Marine Products Board, "About Virginia Seafood."

49   *World's largest source of protein:* United Nations Sustainable Development Knowledge Platform, "Oceans — Facts and Figures."

49   *Overfishing has already resulted:* Froeseller, Kleisner, and Pauly, "What Catch Data Can Tell Us."

49   *Climate disruption may kill off . . . coral reefs:* National Research Council, *Abrupt Impacts,* 112, 120.

6. Tapestry, Interrupted

50   *Single square mile:* Adams and Stephenson, "Twenty-Five Years of Succession," 206.

50   *Covered 700,000 square miles:* Delcourt and Delcourt, "Late Quaternary History," 22–35.

50   *Stable climate, eighty centuries:* Gill interview.

51   *Complex tapestry:* Fleming and Patterson, *Natural Communities of Virginia.*

51   *Can erase this forest:* "As with other mountain systems, the high elevation forests of the southern Appalachian ecosystems are at particular risk from a warming climate. A 3 degree C increase in July temperature would raise climate-elevation bands by about 480 meters, resulting in the extirpation of the rare red spruce–Fraser fir (*Picea rubens* and *Abies fraseri*) alpine forests growing at the highest elevations in North Carolina and harboring federally threatened animal species including the North Carolina flying squirrel" (Ingram, Dow, and Carter, *Southeast Region Technical Report,* 163).

52   *The 15 million acres of other Virginia forests:* Virginia Department of Forestry, "Forest Facts."

52 *Steady upward trend:* Livezey correspondence, data.

52–53 *Threatened, endangered, or shrinking:* Virginia Department of Game and Inland Fisheries, *Virginia's Comprehensive Wildlife Conservation Strategy,* pp. 1–20 to 1–22.

54 *Climate strategy document:* Burkett, *Virginia's Strategy for Safeguarding Species,* 1.

55 *Invasives typically specialize:* Hellman et al., "Five Potential Consequences," 534–43.

55 *Only this brief notation:* McNulty, "George Washington National Forest, Case Study."

55 *Poignantly small, passive steps:* U.S. Forest Service, Southern Region, *Draft Revised Plan,* George Washington National Forest.

56 *Highly likely winners:* Hellman et al., "Five Potential Consequences," 534–43.

56 *Economic value of the ecosystem services produced by Virginia's forests:* Paul, *The Economic Benefits of Natural Goods and Services,* i, ii.

57 *Absorb billions of tons:* Bonan, "Forests and Climate Change," 1444–49.

57 *Old trees also pack on wood growth:* Stephenson et al., "Rate of Tree Carbon Accumulation."

58 *Echo the scenarios:* For example, Dale et al., "Modeling Transient Response," 1888–1901.

58 *"Continued increases in metropolitan populations":* Ingram, Dow, and Carter, *Southeast Region Technical Report,* 161; Neilson interview.

58–59 *Worst insect infestations and fires since Europeans first arrived:* Robbins, "What's Killing the Great Forests?"

59 *"Savannafication":* Ingram, Dow, and Carter, *Southeast Region Technical Report,* 166.

59 *Rhythm of fires may quicken:* Bachelet et al., "VEMAP vs. VINCERA," 38. "[W]armer and drier weather causes the eastern deciduous and mixed forests to shift to a more open canopy woodland or savanna type while boreal forests disappear almost entirely from the Great Lakes area by the end of the 21st century"; and National Science and Technology Council, *Scientific Assessment of the Effects of Global Change on the United States,* 141.

7. Pond and Paradox

60 *Some 13,250 years ago:* Watts, "Late Quaternary Vegetation of Central Appalachia," 427–69. The figures given — 11,140 and 9,140 radiocarbon years before the present — have been converted to calendar years using an updated revision of the carbon dating process, at: http://radiocarbon.ldeo.columbia .edu/cgi-bin/radcarbcal?id=0&fig=1&entry_type=0&add=1&age=11400 &std=100 (Richard G. Fairbanks of Columbia University is the author of the online conversion table.)

60 *Paleovirginians shared the landscape:* McDonald interview.

61 *Material is meticulously combed:* Gill interview.

62 *It will halt many:* Zhu, Woodall, and Clark, "Failure to Migrate," 1051: "Results showed no advantage to potential invaders, certainly not the dominance needed if they were to overcome the numerical disadvantages required for rapid spread. Results of widespread seedling experiments . . . would appear to support model results that predict migration rates far below those required to track contemporary climate change." See also McLachlan, Clark, and Manos, "Molecular Indicators," 2088: "Our estimated rates of 100 m/yr are consistent with model predictions based on life history and dispersal data, and suggest that past migration rates were substantially slower than the rates that will be needed to track 21st-century warming"; and Iverson, Schwartz, and Prasad, "How Fast and Far?," 209: Over a century, "there is a relatively high probability of colonization within a zone of 10–20 km (depending on habitat quality and species abundance) from the current boundary, but a small probability of colonization where the distance from the current boundary exceeds about 20 km."

62 *At a rate of 300 feet to a half mile:* Gill interview.

62 *Even under optimal conditions:* Ibid.; Zhu, Woodall, and Clark, "Failure to Migrate," 1042: "The fact that the majority of seedling extreme latitudes are less than those for adult trees may emphasize the lack of evidence for climate-mediated migration, and should increase concerns for the risks posed by climate change."

62 *Already at or near their northern limit:* Woodall correspondence.

63 *Sinkhole ponds, shale barrens:* Smith correspondence.

66 *Includes some 300,000 acres:* The Nature Conservancy, "Virginia — Places We Protect."

68   *Southern pine beetles in New Jersey Pinelands:* Gillis, "In New Jersey Pines, Trouble Arrives on Six Legs," A1.

68   *Beetles in abeyance:* Johnson interview.

68–69   *Show no signs of migrating north:* Clark interview.

69   *At the edge of the Arctic tundra:* Andreu-Hayles, "Varying Boreal Forest Response."

8. The Fiddlers

71   *Replenishment:* City of Virginia Beach, "Sandbridge Beach Replenishment Fact Sheet 2012/2013."

71   *Along the 10,000 miles of Virginia's tidal shorelines:* Virginia Institute of Marine Science, *Land-Coast Connections.*

72   *Seventy-eight percent of Virginians:* Bernick interview.

72   *The tidal range can vary:* NOAA, "Tides and Currents."

72   *Twice a month, the full moon:* Ibid.

72   *During the tropical storms:* Titus et al., "Virginia," 707.

73   *Measured at more than 9 feet:* Botts et al., *CoreLogic Storm Surge Report 2012,* Virginia sections, p. 19.

73   *Hurricane arrives somewhere in Virginia every 2.3 years:* NOAA, "Virginia Tropical Cyclone Climatology."

73   *Research on more frequent hurricanes inconclusive:* Karl, Melillo, and Peterson, eds., *Global Climate Change Impacts in the United States,* 112: "The destructive potential of Atlantic hurricanes has increased since 1970, correlated with an increase in sea surface temperature. A similar relationship with the frequency of landfalling hurricanes has not been established. . . . The intensity of Atlantic hurricanes is likely to increase during this century with higher peak wind speeds, rainfall intensity, and storm surge height and strength." See also NOAA, National Climate Assessment and Development Advisory Committee, *Draft National Climate Assessment,* 84: "Given the evidence and uncertainties, confidence is medium that the strongest hurricanes are projected to increase in intensity as the oceans warm due to more available energy. Confidence is low regarding other trends in severe storms due to the many uncertainties that remain about frequency and intensity of other types of storms."

73   *Several studies indicate:* NOAA, National Climate Assessment and Development Advisory Committee, *Draft National Climate Assessment,* 112, 388: "An increase in average summer wave heights along the U.S. Atlantic coastline since 1975 has been attributed to a progressive increase in hurricane power." Also: "The intensity of Atlantic hurricanes is likely to increase during this century with higher peak wind speeds, rainfall intensity, and storm surge height and strength."

74   *Each year, the front of that glacier:* Abdalati interview; Shepherd and Ivins, "A Reconciled Estimate," 1183–89.

74   *River of ice:* Joughin correspondence.

74   *"So there's some concern that the sea could rise":* Hansen and Sato, "Paleoclimate Implications," 20.

74   *Adding about 50 cubic miles of water:* Abdalati interview, with gigatons converted to cubic miles.

74   *Antarctica's land-based ice is also melting:* Shepherd and Ivins, "A Reconciled Estimate," 1183–89.

74   *In roughly equal measure:* Bernstein et al., *Assessment Report 4,* Synthesis Report, 30.

76   *Another station has recorded tidal history since 1928:* Stone interview.

76   *About 1 inch every seven or eight years:* Titus and Hudgens, eds., *Likelihood of Shore Protection,* 7.

76   *Some part of that is because coastal land is sinking:* Boon interview.

76   *Has made the land sink even more:* Old Dominion University, Economic Forecasting Project, *The State of the Region — Hampton Roads,* 106.

77   *Would push salt water onto roughly 40 square miles:* Titus and Hudgens, eds., *Likelihood of Shore Protection,* table 8.3, p. 707, halved from the 0–2 feet table.

78   *Ezer led one:* Ezer and Corlett, "Is SLR Accelerating in the Chesapeake Bay?"

78   *Asbury Sallenger:* Sallenger, Doran, and Howd, "Hotspot of Accelerated Sea-Level Rise."

78   *Shows 2 more feet of salt-water rise:* All figures are spring high tide, and use 1992 as the benchmark year for the comparisons. Data at http://corpsclimate.us/ccaceslcurves.cfm.

78   *"A cautious and conservative approach":* Pilkey and Young, *Rising Sea,* 79.

78 *Cites 7.5 feet of sea level rise by around 2100:* Mitchell et al., *Recurrent Flooding Study*, 78.

78–79 *They will continue for an indefinite period:* Bernstein et al., *Assessment Report 4*, Synthesis Report, 47.

79 *U.S. Geological Survey has yet to study:* Bennett interview.

84 *Could inundate 82 square miles of dry land:* Titus and Hudgens, eds., *Likelihood of Shore Protection*, table 8.3, p. 706. The figures exclude Charles City, Chesterfield, Colonial Heights, Franklin, Hanover, Henrico, Hopewell, New Kent, Petersburg, Prince George, Southampton, and Williamsburg Counties, which this study did not examine.

84 *Fifteen miles of Virginia interstate, and details that follow:* Titus et al., *Coastal Sensitivity*, table 7.8, p. 363.

84 *Because transportation systems are networks:* Ibid., 364.

84 *The Hampton Roads Planning District Commission estimated:* McFarlane and Hampton Roads Planning District Commission, *Climate Change in Hampton Roads*, table 6, p. 44.

85 *As many as 628,000 Virginians:* Titus et al., *Coastal Sensitivity*, draft version, table F.6, pp. 691–96.

85 *"Trees, autos, boats":* Botts et al., *Corelogic Storm Surge Report 2012*, Virginia sections, 7.

85 *"This time the other people":* Wieseltier, "Who by Water," 52.

9. Poker

87 *Ranks Virginia Beach second only to New York:* Botts, *Corelogic Storm Surge Report 2012*, Virginia sections, 19. Its research estimates that surge from a Category 1 hurricane would damage more than 59,000 residences in Virginia Beach today at a cost of $10 billion. A Category 5 hurricane would cause $46 billion in damage to 290,000 homes.

87 *One of the top twenty in the world:* Nicholls et al., "Organization for Economic Cooperation and Development Ranking," table 2, p. 4.

88 *Virginia climate change commission's recommendations for planning for sea level rise:* Governor's Commission on Climate Change, *Final Report*, 36, recommendation 14K.

88   *Life span of eighty or ninety years:* Bernick interview.

88   *No guidance from Virginia Department of Transportation:* Rollison interview.

89   *The cost of seawalls:* Pilkey and Young, *Rising Sea,* 170.

91   *"Storm surge, storm waves":* Ibid., 178.

91   *Paul Fraim had the courage to say:* Brangham, "The Rising Seas."

95   *Yet another haunting competitive ranking:* Institute for Environmental Negotiation, *Sea Level Rise in Hampton Roads,* 9.

95   *Seventy-eight miles upstream:* Port of Richmond, "City of Richmond Marine Facility Lessee."

96   *"Urban development has increased":* National Capital Planning Commission, *Report on Flooding,* 1.

96   *A new $10 million pop-up floodwall:* www.nab.usace.army.mil/Media /FactSheets/FactSheetArticleView/tabid/10470/Article/8411/washington -dc-vicinity-local-flood-protection.aspx.

96   *What would now be tens of millions of dollars in damage:* U.S. Department of Labor, "CPI Inflation Calculator."

98   *"Affected citizens will pressure local government":* Boon, Wang, and Shen, *Planning for Sea Level Rise and Coastal Flooding,* 4.

98–99   *Until the latest $15 million round:* City of Virginia Beach, "Sandbridge Beach Replenishment Fact Sheet 2012/2013."

99   *Army Corps of Engineers outlays:* Pawlik correspondence; Federal Emergency Management Agency, "Policy Statistics Virginia as of 11/30/2012."

99   *Called forth another $51 billion in federal help:* Chen, "U.S. to Release First Installment of $51 Billion in Hurricane Sandy Aid," A22.

10. Entrepreneurs

101   *The climate cannot stabilize for a century or so:* National Research Council, *Climate Stabilization Targets,* 5.

103   *Data from 3 million years ago:* Robinson interview.

103   *Its geographic signature:* Krantz, "A Chronology of Pliocene Sea-Level Fluctuations," 165.

104 *Most of those natural areas will drown:* Bilkovic et al., *Vulnerability of Shallow Tidal Habitats,* v.

104 *One study lists twenty-five different species:* Street et al., *North Carolina Coastal Habitat Protection Plan.*

104 *"Significant loss":* Pyke et al., *Climate Change and the Chesapeake Bay,* 40.

104 *For nursery and spawning grounds:* Ibid.

104 *About 660 square miles of shallow water:* Titus and Hudgens, eds., *Likelihood of Shore Protection,* table 8.3, p. 706. The figures exclude Charles City, Chesterfield, Colonial Heights, Franklin, Hanover, Henrico, Hopewell, New Kent, Petersburg, Prince George, Southampton, and Williamsburg Counties, which this study did not examine.

104 *Given the sea level rise that most recent studies anticipate:* Bilkovic et al., *Vulnerability of Shallow Tidal Water Habitats,* 16, 25.

105 *Wetlands will not survive the century:* for example, Titus et al., *Coastal Sensitivity,* 29.

105 *All eelgrass beds could disappear:* Bilkovic et al., *Vulnerability of Shallow Tidal Habitats,* 16, 25, 30, 39.

105 *VIMS tallies have found:* Ibid., 6; VIMS, Center for Coastal Resources Management, *Regulatory Fidelity to Guidance,* fig. 5, p. 17.

106 *"If we're going to change society":* Rogers, "People, Planet, Profit," 36–37.

106 *He described Bluff Point:* Ibid.

106 *A 900-acre spread:* Bluff Point Holdings LLC, Bluff Point Virginia Master Plan Overview.

106 *"A good time for all of us to ask":* Titus et al., "State and Local Governments Plan," 6.

107 *Already built upon:* Ibid.

107 *"Few institutional limitations":* Ibid., 4.

108 *Often referred to as "rolling easements":* Titus and U.S. Environmental Protection Agency, *Climate Ready Estuaries Program,* 4.

111–112 *Currently reaps about $42 million:* Hinds correspondence.

112   *For the region, the figure is $150 million:* Ibid.

112   *Replenishing the sand barrier would cost $24 million:* Hinds interview.

11. Resettlements

115   *Hawks and hummingbirds, mosquito species and their active seasons:* Gaines interview.

115   *Came across an odd book:* Delcourt and Delcourt, *Living Well in the Age of Global Warming.*

116   *This field of research — migration studies:* See, for example, Lueck, "Environment Migration," 2; and Hopkins interview.

117   *Ethnic-minority, low-income, low-education, and female-headed households:* Lueck, "Environment Migration," 2.

117   *Rivaled the Dust Bowl:* Grier, "The Great Katrina Migration."

117   *"A sea of antagonism":* Stein, *California and the Dust Bowl Migration,* 45, 79; Grier, "The Great Katrina Migration."

117   *"Bum blockade":* Stein, *California and the Dust Bowl Migration,* 73–74.

117   *"California is faced with economic chaos":* Gregory, *American Exodus,* 95.

117   *"The scale is monumental":* Grier, "The Great Katrina Migration."

117   *The official count was that 3,300 Katrina refugees:* McIntosh, "Measuring the Labor Market Impacts," 21.

118   *"The majority of the people here":* Hopkins, "Flooded Communities," 1–2.

118   *"We didn't know white people could love us":* Ibid.

118   *"Migrations transport place-specific disasters to other communities":* Crowley, "Where Is Home?" 121–66, cited in Lueck, "Environment Migration."

118   *Illegal immigrants were eligible:* Lueck, "Environment Migration," 8.

119   That crime-is-the-biggest-problem number rose: Hopkins, "Flooded Communities," 12.

119   *"Paying more attention to fewer crimes":* Ibid.

120   *Perhaps scientists and policy makers can identify policies:* Fussell, "Disasters and Migration."

120 *"The fate of the poor":* Karl, Melillo, and Peterson, eds., *Global Climate Change Impacts in the United States,* 101.

121 *List of potential threats:* Gaines interview.

121 *Reported Lyme cases:* Virginia Department of Health, "Table 2: Ten-Year Trend."

121 *More accurate number of infections would probably be 10,000:* CDC, "CDC Provides Estimate," 1.

122 *About 60 new imported cases of malaria*: Virginia Department of Health "Table 2: Ten-Year Trend."

122 *Spielman quotes:* Mann, *1493, 113.*

122 *CDC created to fight malaria:* CDC, "Our History."

122 *Kern County, California:* Reisen et al., "Delinquent Mortgages."

125 *"People who want to give a group hug":* Revkin, "On Issues Like Climate Change."

12. Virginia Climate Fever: R$_x$

127 *The rescue plans differ, but many converge:* See, for example, Randers and Gilding, "The One Degree War Plan," 170–88; the IPCC strategy using "carbon wedges," as summarized in Kitchen, *Global Climate Change,* 331– 39; and Archer, *The Long Thaw,* "Epilogue: Carbon Economics and Ethics."

128 *"Conservation may be a sign of personal virtue":* Kahn, "Cheney, on the Road."

129 *A quarter of the state's river and stream miles:* VIRGINIA Draft 305(b)/303(d) Water Quality Integrated Report to Congress, March 2012.

130 *Highly visible citizens' groups in Virginia:* The Chesapeake Climate Action Network, the Citizens Climate Lobby, and many of the members of the Virginia Conservation Network that have the climate issue at or near the top of their agendas are easy to find.

130 *Three recommendations:* Shabecoff, "E.P.A. Proposes Rules to Curb Warming."

130 *We are still 15 miles per gallon short:* National Highway Traffic Safety Administration, "Summary of Fuel Economy Performance," unpaged.

131 *"Critical Infrastructure Protection and Resiliency Strategic Plan":* Commonwealth of Virginia, *Critical Infrastructure Protection and Resiliency Strategic Plan.*

133 *No single agency has tracked their fate:* Hayden correspondence; Paylor correspondence.

134 *Texas solar workers:* Hargreaves, "Solar Jobs Outnumber Ranchers."

134 *Major features of Maryland's plan:* Maryland Department of the Environment, "Executive Summary," 6, 7.

134 *Virginia is one of nineteen states:* U.S. Department of Energy, "Most States Have Renewable Portfolio Standards."

135 *We rank far behind other states:* Marine Conservation Institute, "How Well Does Your State Protect Your Coastal Waters?" 5.

137–38 *Global campaign needed to address climate change:* Randers and Gilding, "The One Degree War Plan," 170–88.

13. Oracles

139 *A quadrillion-calculations-per-second thinker:* National Center for Atmospheric Research, "NCAR Selects IBM," 1.

139 *Yellowstone has thirty times the computing power:* Ibid.

139 *Cost about $30 million and the following details:* Loft interview.

139 *"Full fat tree":* Loft interview.

139 *Fifth-fastest computer in the world:* Ibid.

139 *That much electrical energy could power:* Loft's calculation, based in part on www.eia.gov/tools/faqs/faq.cfm?id=97&t=3.

139 *Which explains Cheyenne:* Loft interview.

139 *Just one of those commonplace ironies:* Institute for Energy Research, "Wyoming Energy Facts."

140 *Complex coastlines and mountain ranges:* Meehl interview.

140 *Uses roughly fifty times more computing power:* Holland interview.

140 *Grid units as small as 7 miles on a side:* Ibid.

140 *Occoquan Reservoir:* Fairfax Water, "About."

141 *"A kind of prosthetic God":* Gay, *Freud: A Life,* 546.

141 *"Essentially, all models are wrong":* Box and Draper, *Empirical Model-Building,* 424.

144 *Clouds:* Holland interview.

144 *Recent studies point to the causes of this warming hole:* See the explanation, for instance, in Meehl, Arblaster, and Branstator, "Mechanisms Contributing to the Warming Hole."

145 *Sickens the lungs and hearts:* Environmental Protection Agency, "Overview of EPA's Revisions to Air Quality Standards for Particle Pollution (Particulate Matter)," 1; see also Nash, *Blue Ridge 2020,* 61–76.

145 *Poisons mountain soils:* Nash, *Blue Ridge 2020,* 51–60.

145 *Sulfate particles may have shielded Virginia:* Leibensperger et al., "Climatic Effects of 1950–2050 Changes," 3333–48.

146 *The earth's overall average surface temperature has not warmed much, and discussion:* Tollefson, "The Case of the Missing Heat," 276–78; see also Meehl et al., "Model-Based Evidence of Deep-Ocean Heat Uptake"; Watanabe et al., "Strengthening of Ocean Heat Uptake Efficiency"; and Cowtan and Way, "Coverage Bias."

148 *Climate disruption may invalidate that assumption:* Terando interview; Robinson interview; see also Dalton and Jones, comps., *Southeast Regional Assessment Project,* 36: "The primary disadvantage of statistical downscaling is the assumption of stationarity in the predictor-predictand relation, which assumes little to no change in the climate system feedback mechanisms through time (Wilby, 1998)."

148 *Climate models cannot contemplate "unknown unknowns":* Schmidt, "How Earth System Models Are Reshaping the Science/Policy Interface."

148 *"Some changes may occur in a relatively predictable way":* NOAA, National Climate Assessment and Development Advisory Committee, *Draft National Climate Assessment,* 13.

148 *Best predictor:* Schmidt, "How Earth System Models Are Reshaping the Science/Policy Interface."

14. A Hierarchy of Credibility

151   *Questions about climate science are answered by research:* Pielke, *The Honest Broker.*

153   *Of the 928 climate research papers:* Oreskes, "Scientific Consensus," 1686.

153   *In another study, 3,146 geochemists:* Doran and Zimmerman, "Examining the Scientific Consensus," 22–23.

154   *Of 4,013 climate research papers:* Cook et al., "Quantifying the Consensus," 1–7.

154   *Then 1,372 climate researchers:* Anderegga et al., "Expert Credibility," 1–3.

154   *The state climatologist for twenty-seven years:* Curriculum vitae supplied by Patrick J. Michaels.

155   *Written by a former television weatherman:* Watts, "About" and "Publications and Projects."

159   *Confirmed in nearly endless detail:* Oreskes and Conway, *Merchants of Doubt.*

159   *Concealed the sources of their funding even more actively:* Brulle, "Institutionalizing Delay."

159   *Michaels declined to reveal the names of clients:* See Michaels's motion-to-intervene affidavit in U.S. District Court for the District of Vermont, Green Mountain Chrysler Plymouth Dodge Jeep, et al. vs. George Crombie, filed July 6, 2007, pp. 1–4.

   Michaels's written comment to me regarding the affidavit: "My major issue is that no environmental scientist I know lists their consulting contracts on their CV. It is considered bad form." In a later message, he added: "I think you should go to the website for the Environmental Sciences Department at U.Va. (my old department) and have a look at the CVs of the faculty. I don't know one that listed consulting income—but you should see for yourself. That is the authority that governed the form of my CV."

   But I contacted Patricia Wiberg, the head of that department, and this was her reply: "I can only speak for the faculty in our department and any other CVs I recall seeing in other contexts but I'd say that statement is on the strong side. First, at any given time relatively few of our faculty are engaged in any significant consulting activities. It is probably true that there are some people doing small consulting jobs who don't list them on their

CVs because they represent such a small part of their time and effort. However, a quick search of the 37 CVs that were submitted to me as part of last year's annual faculty review process in our department (tenured, tenure-track and research faculty) showed that 15 people list some paid consulting, now or in the past, on their CVs — about 40% of the faculty."

160 *Zakaria discussion, questions:* Cato Institute, "Patrick J. Michaels Discusses," located at 6:45 on video.

161–62 *Rear Admiral David Titley:* Titley, Blue Planet Forum lecture, Old Dominion University, December 2, 2010.

163 *Snowe, Rockefeller to Exxon:* Sandell, "Senators to Exxon: Stop the Denial."

164 *Became well known for her behavioral model:* Though by no means is the model always agreed with in the research literature.

BIBLIOGRAPHY

Interviews and Correspondence

Abdalati, Waleed. Glaciologist, Earth Science and Observation Center, University of Colorado.

Akerlof, Karen. Research assistant professor, Center for Climate Change Communication, George Mason University.

Anderson, Mark. Director, conservation science, The Nature Conservancy.

Arndt, Derek. Chief, Climate Monitoring Branch, National Climatic Data Center.

Arnold, Tom. Chemical ecologist, Dickinson College.

Bachelet, Dominique. Senior climate change scientist, Conservation Biology Institute.

Beard, Ben. Associate director for climate change; chief, Bacterial Diseases Branch, Division of Vector-Borne Diseases, Centers for Disease Control and Prevention.

Bennett, Mark. Director, USGS Virginia Water Science Center, Richmond.

Bernick, Clay. Administrator, Environmental Sustainability Office, Virginia Beach.

Boon, John. Emeritus professor, Virginia Institute of Marine Science.

Breitburg, Denise. Marine ecologist, Smithsonian Environmental Research Center.

Brooke, Sandra. Marine biologist, senior conservation fellow, Marine Conservation Institute.

Burkett, Chris. Wildlife Action Plan coordinator, Department of Game and Inland Fisheries.

Butler, James H. Director, Global Monitoring Division, Earth System Research Laboratory, National Oceanic and Atmospheric Administration.

Chaytor, Jason. Geologist, Woods Hole Oceanographic Institution.

Clark, James. Forest ecologist, Duke University.

Cogan, Jonathan. Media relations, U.S. Energy Information Administration.

Dingledine, Thomas. Real estate developer.

Dowsett, Harry. Geologist, U.S. Geological Survey.

Duhring, Karen. Coastal resource management scientist, Virginia Institute of Marine Science.

Erwin, Michael. Wildlife biologist, U.S. Geological Survey.

Gaines, David. Virginia state public health entomologist.

Gill, Jacquelyn. Paleoecologist and biogeographer, Brown University.

Hayden, Bill. Public Affairs Office, Virginia Department of Environmental Quality.

Hallerman, Eric. Fisheries biologist, Virginia Polytechnic Institute and State University.

Hayhoe, Katharine. Climate scientist; director, Climate Science Center, Texas Tech University.

Hershner, Carl. Director, Center for Coastal Resources Management, Virginia Institute of Marine Science.

Hinds, Louis. Retired manager, Chincoteague National Wildlife Refuge.

Holland, Marika. Chief scientist, Community Earth System Modeling Project, National Center for Atmospheric Research.

Hönisch, Bärbel. Paleo-oceanographer, Lamont-Doherty Earth Observatory, Columbia University.

Hopkins, Daniel. Political scientist, Georgetown University.

Johnson, Derek M. Biologist, Virginia Commonwealth University.

Joughin, Ian. Principal investigator, Polar Science Center, University of Washington.

Kesterson, Julian. Observer, National Weather Service.

Kunkel, Kenneth. Senior climate scientist, National Oceanic and Atmospheric Administration.

Landgraf, Kenneth. Supervising planner, George Washington National Forest.

Livezey, Robert. Consultant statistician; chief of Climate Services, National Weather Service, retired.

Loft, Richard. Director for technology development, Computational and Information Systems Laboratory, National Center for Atmospheric Research.

Madron, Justin. Geographic information system analyst, University of Richmond.

McDonald, Jerry. Geographer, archaeologist, and publisher, McDonald and Woodward Publishing Company.

McNulty, Steve. Team leader, Southern Climate Change Research, U.S. Forest Service.

Meehl, Gerald. Senior scientist, National Center for Atmospheric Research.

Michaels, Patrick. Director, Center for the Study of Science, Cato Institute.

Miller, Whitman. Marine ecologist, Smithsonian Environmental Research Center.

Neilson, Ron. Bioclimatologist, U.S. Forest Service.

Noss, Reed. Conservation biologist, University of Central Florida.

Nuccitelli, Dana. Contributor, Skeptical Science website. www.skepticalscience
.com.

Nye, Janet. Fisheries ecologist, State University of New York at Stony Brook.

Olson, David. Spokesman, U.S. Army Corps of Engineers.

Ostfeld, Richard. Disease ecologist, Cary Institute of Ecosystem Studies.

Pawlik, Eugene. Spokesman, U.S. Army Corps of Engineers.

Paylor, David K. Director, Virginia Department of Environmental Quality.

Robinson, Marci. Geologist, U.S. Geological Survey.

Rollison, Tamara. Spokesperson, Virginia Department of Transportation.

Ross, Steve. Fisheries biologist, University of North Carolina, Wilmington.

Shugart, Herman K. (Hank). Forest ecologist, University of Virginia.

Simpson, Cyndi. Minister, Unitarian Church of Norfolk.

Smith, Tom. Director, Natural Heritage, Virginia Department of Conservation
and Recreation.

Stenger, Phillip J. (Jerry). Director, Office of Climatology, University of Virginia.

Stiles, Skip. Executive director, Wetlands Watch.

Stone, Peter. Chief, Oceanographic Division, National Oceanic and Atmospheric
Administration, Center for Operational Oceanographic Products and Services,
Silver Spring, Md.

Sullivan, Robert. Spokesperson, CSX Corporation.

Terando, Adam. Climate change research coordinator, North Carolina State
University.

Titus, James G. Sea-level rise manager, U.S. Environmental Protection Agency.

Vincent, Katy. Communications coordinator, National Climatic Data Center.

Waldbusser, George. Ocean ecologist and biogeochemist, Oregon State
University.

Weisskohl, Marjorie. Spokesperson, Bureau of Ocean Energy Management.

Woodall, Christopher. U.S. Forest Service, forest inventory and analysis specialist.

Books and Articles

Adams, Harold S., and Steven L. Stephenson. "Twenty-Five Years of Succession in
the Spruce-Fir Forest on Mount Rogers in Southwestern Virginia." *Castanea* 75,
no. 2 (June 2010): 205–10.

Anderegga, William R. L., James W. Prallb, Jacob Harold, and Stephen H. Schnei-
der. "Expert Credibility in Climate Change." *Proceedings of the National Academy of Sciences* 107, no. 27 (April 2010): 1–3.

Andreu-Hayles, Laia, Rosanne D'Arrigo, Kevin J Anchukaitis, Pieter S. A. Beck,

David Frank, and Scott Goetz. "Varying Boreal Forest Response to Arctic Environmental Change at the Firth River, Alaska." *Environmental Research Letters* 6 (October 2011). iopscience.iop.org/1748-9326/6/4/045503/fulltext/.

Appalachian Voices, Virginia Chapter of the Sierra Club, and Chesapeake Climate Action Network. *Dirty Money, Dirty Power—How Virginia's Energy Policy Serves the Interests of Top Campaign Contributors.* October 2, 2012. appvoices.org/images/uploads/2012/11/Dirty-Energy-Politics.pdf.

Archer, D. *The Long Thaw: How Humans Are Changing the Next 100,000 Years of Earth's Climate.* Princeton, N.J.: Princeton University Press, 2008.

Arnold, T., Christopher Mealey, Hannah Leahey, A. Whitman Miller, Jason M. Hall-Spencer, Marco Milazzo, and Kelly Maers. "Ocean Acidification and the Loss of Phenolic Substances in Marine Plants." *PLOS ONE* 7, no. 4 (2012): e35107. plosone.org/article/info%3Adoi%2F10.1371%2Fjournal.pone.0035107.

Bachelet, D., J. Lenihan, R. Drapek, and R. Neilson. "VEMAP vs VINCERA: A DGVM Sensitivity to Differences in Climate Scenarios." *Global and Planetary Change* 64 (2008): 38–48.

Berger, A., and M. F. Loutre. "Insolation Values for the Climate of the Last 10 Million Years." *Quaternary Science Reviews* 10 (1991): 297–317.

Bernstein, Lenny, Peter Bosch, Osvaldo Canziani, Zhenlin Chen, Renate Christ, Ogunlade Davidson, William Hare, et al. Intergovernmental Panel on Climate Change. *Assessment Report 4.* 2007.

Betts, Richard A., Matthew Collins, Deborah L. Hemming, Chris D. Jones, Jason A. Lowe, and Michael Sanderson. *When Could Global Warming Reach 4°C?* Hadley Centre Technical Note 80. December 2009. Unpaginated.

Bilkovic, Donna Marie, Carl Hershner, Tamia Rudnicky, Karinna Nunez, Dan Schatt, Sharon Killeen, and Marcia Berman. *Vulnerability of Shallow Tidal Water Habitats in Virginia to Climate Change.* Center for Coastal Resources Management, Virginia Institute of Marine Science. 2009.

Bluff Point Holdings, LLC. Bluff Point Virginia Master Plan Overview. http://bluffpointva.com/pdf/MasterPlanOverview.pdf.

Bonan, G. B. "Forests and Climate Change: Forcings, Feedbacks, and the Climate Benefits of Forests." *Science* 320 (2008): 1444–49.

Boon, John, Harry Wang, and Jian Shen. "Planning for Sea Level Rise and Coastal Flooding." Virginia Institute of Marine Science. October 2008.

Boon, John D., John M. Brubaker, and David R. Forrest. *Chesapeake Bay Land Subsidence and Sea Level Change: An Evaluation of Past and Present Trends and Future Outlook. A Report to the U.S. Army Corps of Engineers Norfolk District.* Special Report No. 425 in Applied Marine Science and Ocean Engineering. Virginia Institute of Marine Science. November 2010.

Boon, John D., Harry Wang, and Jian Shen. "Planning for Sea Level Rise and Coastal Flooding." Virginia Institute of Marine Science. October 2008.

Botts, Howard, Thomas Jeffery, Steven Kolk, and Logan Suhr. *CoreLogic Storm Surge Report 2012.* CoreLogic, Inc.

Box, George E. P., and Norman R. Draper. *Empirical Model-Building and Response Surfaces.* Hoboken: Wiley, 1987.

Brangham, William. "The Rising Seas: Mayor of Norfolk, Virginia, Warns That Parts of the City May Need to Be Abandoned." *Atlantic Online,* April 30, 2012. www.theatlantic.com/technology/archive/2012/04/the-rising-seas-mayor-of -norfolk-virginia-warns-that-parts-of-the-city-may-need-to-be-abandoned /256515/.

Brody, Jane E. "When Lyme Disease Lasts and Lasts." *New York Times,* July 8, 2013.

Brulle, Robert J. "Institutionalizing Delay: Foundation Funding and the Creation of U.S. Climate Change Counter-Movement Organizations." *Climatic Change,* December 2013.

Burkett, C. *Virginia's Strategy for Safeguarding Species of Greatest Conservation Need from the Effects of Climate Change.* Virginia Department of Game and Inland Fisheries. 2009.

Caldeira, K., and M. E. Wickett. "Ocean Model Predictions of Chemistry Changes from Carbon Dioxide Emissions to the Atmosphere and Ocean." *Journal of Geophysical Research* 110 (2005): 1–12.

Cato Institute. "Patrick J. Michaels Discusses the Environment and Climate Change on CNN's GPS with Fareed Zakaria." www.cato.org/multimedia/video -highlights/patrick-j-michaels-discusses-environment-climate-change-cnns -gps-fareed-zakaria.

Centers for Disease Control and Prevention (CDC). "CDC Provides Estimate of Americans Diagnosed with Lyme Disease Each Year." Press release. August 19, 2013. www.cdc.gov/media/releases/2013/p0819-lyme-disease.html.

———. *Extreme Heat: A Prevention Guide to Promote Your Personal Health and Safety.* 2006. www.bt.cdc.gov/disasters/extremeheat/heat_guide.asp.

———. "Our History — Our Story." www.cdc.gov/about/history/ourstory.htm.

Chen, David W. "U.S. to Release First Installment of $51 Billion in Hurricane Sandy Aid." *New York Times,* February 5, 2013.

City of Virginia Beach, Virginia. "Sandbridge Beach Replenishment Fact Sheet 2012/2013."

Commonwealth of Virginia. Office of Commonwealth Preparedness. *Critical Infrastructure Protection and Resiliency Strategic Plan.* May 3, 2008.

Cook, John, Dana Nuccitelli, Sarah A. Green, Mark Richardson, Bärbel Winkler, Rob Painting, Robert Way, Peter Jacobs, and Andrew Skuce. "Quantifying the Consensus on Anthropogenic Global Warming in the Scientific Literature." *Environmental Research Letters* 8, no. 2 (January 2013): 1–7.

Cowtan, Kevin, and Robert G. Way. "Coverage Bias in the HadCRUT4 Temperature Series and Its Impact on Recent Temperature Trends." *Quarterly Journal of the Royal Meteorological Society.* Prepublication version. Unpaginated.

Crowley, S. "Where Is Home? Housing for Low-Income People after the 2005 Hurricanes." In *There Is No Such Thing as a Natural Disaster: Race, Class, and Hurricane Katrina,* edited by C. Hartman and G. D. Squires, 121–66. New York: Routledge, 2006.

Cui, Ying, Lee R. Kump, Andy J. Ridgwell, Adam J. Charles, Christopher K. Junium, Aaron F. Diefendorf, Katherine H. Freeman, Nathan M. Urban, and Ian C. Harding. "Slow Release of Fossil Carbon during the Palaeocene-Eocene Thermal Maximum." *Nature Geoscience* 4 (June 2011): 481–85.

Dale, Virginia H., M. Lynn Tharp, Karen O. Lannom, and Donald G. Hodges. "Modeling Transient Response of Forests to Climate Change." *Science of the Total Environment* 408 (2010): 1888–901.

Dalton, M. S., and S. A. Jones, comps. *Southeast Regional Assessment Project for the National Climate Change and Wildlife Science Center. U.S. Geological Survey Open-File Report #2010–1213.* 2010.

Delcourt, Hazel R., and Paul A. Delcourt. "Late Quaternary History of the Spruce-Fir Ecosystem in the Southern Appalachian Mountain Region." In *The Southern Appalachian Spruce-Fir Ecosystem: Its Biology and Threats — Uplands Field Research Station Report,* edited by Peter S. White, 22–35. 1984.

Delcourt, Paul, and Hazel Delcourt. *Living Well in the Age of Global Warming — Ten Strategies for Boomers, Bobos, and Cultural Creatives.* White River Junction, Vt.: Chelsea Green, 2001.

Doran, Peter T., and Maggie Kendall Zimmerman. "Examining the Scientific Consensus on Climate Change." *EOS Transactions* 90, no. 3 (January 20, 2009): 22, 23.

Dowsett, H. J., M. M. Robinson, D. K. Stoll, and K. M. Foley. "Mid-Piacenzian Mean Annual Sea Surface Temperature Analysis for Data-Model Comparisons." *Stratigraphy* 7 (2010): 189–98.

Dyke, A. S., and V. K. Prest. "Late Wisconsinan and Holocene History of the Laurentide Ice Sheet." *Geographie Physique et Quaternaire* 41 (1987): 237–64.

Ezer, Tal, and William Bryce Corlett. "Is SLR Accelerating in the Chesapeake Bay? A Demonstration of a Novel New Approach for Analyzing Sea Level Data." *Geophysical Research Letters* 39 (October 4, 2012).

Fairfax Water. "About." www.fcwa.org/.

Fall, Souleymane, Anthony Watts, John Nielsen-Gammon, Evan Jones, Dev Niyogi, John R. Christy, and Roger A. Pielke Sr. "Analysis of the Impacts of Station Exposure on the U.S. Historical Climatology Network Temperatures and Temperature Trends." *Journal of Geophysical Research* 116 (July 27, 2011): D14.

Federal Emergency Management Agency. "Policy Statistics Virginia as of 11/30/2012." http://bsa.nfipstat.fema.gov/reports/1011.htm#VAT.

Field, C. B., et al. *Climate Change 2007: Impacts, Adaptation and Vulnerability. Contribution of Working Group II to the Fourth Assessment Report of the Intergovernmental Panel on Climate Change.* Cambridge: Cambridge University Press, 2007.

Fleming, Gary P., and Karen D. Patterson. *Natural Communities of Virginia: Ecological Groups and Community Types.* Natural Heritage Technical Report 12–04. Virginia Department of Conservation and Recreation, Division of Natural Heritage, Richmond. 2012.

Friedland, Kevin D., Joe Kane, Jonathan A. Hare, R. Gregory Lough, Paula S. Fratantoni, Michael J. Fogarty, and Janet A. Nye. "Thermal Habitat Constraints on Zooplankton Species Associated with Atlantic cod (Gadus morhua) on the US Northeast Continental Shelf." *Progress in Oceanography* 116 (September 2013): 1–13.

Froeseller, Rainer, K. Kleisner, and D. Pauly. "What Catch Data Can Tell Us about the Status of Global Fisheries." *Marine Biology,* March 9, 2012, 1283–92.

Fussell, Elizabeth. "Disasters and Migration." The Encyclopedia of Global Human Migration. February 2013. http://onlinelibrary.wiley.com/doi/10.1002/9781444351071.wbeghm175/abstract.

Fussell, E., N. Sastry, and M. VanLandingham. "Race, Socioeconomic Status, and Return Migration to New Orleans after Hurricane Katrina." *Population & Environment* 31, no. 1 (2010): 20–42.

Gay, Peter. *Freud: A Life for Our Time.* Reprint, New York: Norton, 1988.

Gillis, Justin. "In New Jersey Pines, Trouble Arrives on Six Legs." *New York Times,* December 1, 2013, A1.

Governor's Commission on Climate Change, Commonwealth of Virginia. *Final Report: A Climate Change Action Plan.* December 15, 2008.

Gregory, James N. *American Exodus.* New York: Oxford University Press, 1989.

Grier, Peter. "The Great Katrina Migration." *Christian Science Monitor,* September 12, 2005. www.csmonitor.com/2005/0912/p01s01–ussc.html.

Hansen, James E., and Makiko Sato. "Paleoclimate Implications for Human-Made Climate Change." In *Climate Change,* edited by A. Berger, F. Mesinger, and D. Šijački. Vienna: Springer-Verlag, 2012.

Hargreaves, Steve. "Solar Jobs Outnumber Ranchers in Texas, Actors in California." CNN Money, April 22, 2013. http://money.cnn.com/2013/04/22/news/economy/solar-jobs/index.html.

Hayden, Bruce P., and Patrick J. Michaels, "Virginia's Climate." University of Virginia Climatology Office. http://climate.virginia.edu/description.htm.

Hayhoe, Katharine. Interview on *Nova Science Now.* www.pbs.org/wgbh/nova
/secretlife/scientists/katharine-hayhoe/.

Hayhoe, Katharine, and Andrew Farley. *A Climate for Change.* New York: Faith
Words, 2011.

Hellmann, Jessica J., James E. Byers, Britta G. Bierwagen, and Jeffrey S. Dukes.
"Five Potential Consequences of Climate Change for Invasive Species." *Conservation Biology* 22, no. 3 (June 2008): 534–43.

Hill, Steven. "Windmills, Tides, and Solar Besides: The European Way of Energy,
Transportation, and Low-Carbon Emissions." *Environmental Law Reporter* 43,
no. 2 (February 2013): 10102–20.

Holdren, John. "Meeting the Climate-Change Challenge." National Council for
Science and the Environment. Eighth Annual John H. Chafee Memorial Lecture on Science and the Environment. January 17, 2008.

Hönisch, B., Andy Ridgwell, Daniela N. Schmidt, Ellen Thomas, Samantha J.
Gibbs, Appy Sluijs, Richard Zeebe, et al. "The Geological Record of Ocean
Acidification." *Science* 335, no. 6072 (2012): 1058–63.

Hopkins, Daniel J. "Flooded Communities: Explaining Local Reactions to the
Post-Katrina Migrants." *Political Research Quarterly* 65, no. 2 (November
2009): 443–59.

Ingram, Keith T., Kirstin Dow, and Lynne Carter. *Southeast Region Technical
Report to the National Climate Assessment, 2012.*

Institute for Energy Research, "Wyoming Energy Facts." 2012. www.institutefor
energyresearch.org/state-regs/pdf/Wyoming.pdf.

Institute for Environmental Negotiation, University of Virginia, Virginia Sea
Grant Project. *Sea Level Rise in Hampton Roads: Findings from the Virginia
Beach Listening Sessions.* March 30–31, 2011.

Intergovernmental Panel on Climate Change. *Summary for Policymakers —
Emissions Scenarios. A Special Report of IPCC Working Group III.* 2000.

International Atomic Energy Agency. "World Inter-regional Hard Coal Net
Trade." *World Energy Outlook, by Major Region, 2011.*

International Energy Agency. "CO_2 Emissions from Fuel Combustion Highlights." 2012. www.iea.org/newsroomandevents/news/2012/may/name,27216
,en.html.

———. "Global Carbon-Dioxide Emissions Increase by 1.0 Gt in 2011 to Record
High." Press release. May 2012. www.iea.org/newsroomandevents/news/2012
/may/name,27216,en.html.

———. *Medium-Term Coal Market Report 2013, Executive Summary,* 11–13.

Iverson, Louis R., M. W. Schwartz, and Anantha M. Prasad. "How Fast and
Far Might Tree Species Migrate in the Eastern United States Due to Climate
Change?" *Global Ecology and Biogeography* 13 (2004): 209–19.

Jefferson, Thomas. *Notes on the State of Virginia*. 1781. Electronic Text Center, University of Virginia Library. http://etext.virginia.edu/etcbin/toccernew2 ?id=JefVirg.sgm&images=images/modeng&data=/texts/english/modeng /parsed&tag=public&part=7&division=div1.

Kahn, Joseph. "Cheney, on the Road, Seeks Support for Energy Program." *New York Times*, July 17, 2001.

Karl, Thomas R., Jerry M. Melillo, and Thomas C. Peterson, eds. *Global Climate Change Impacts in the United States*. Cambridge: Cambridge University Press, 2009.

Kitchen, David. *Global Climate Change — Turning Knowledge into Action*. Boston: Pearson Education, 2014.

Krantz, David E. "A Chronology of Pliocene Sea-Level Fluctuations: The U.S. Middle Atlantic Coastal Plain Record." *Quaternary Science Reviews* 10 (1991): 163–74.

Leibensperger, E. M., L. J. Mickley, D. J. Jacob, W.-T. Chen, J. H. Seinfeld, A. Nenes, P. J. Adams, D. G. Streets, N. Kumar, and D. Rind. "Climatic Effects of 1950–2050 Changes in US Anthropogenic Aerosols — Part 1: Aerosol Trends and Radiative Forcing." *Atmospheric Chemistry and Physics* 12 (2012): 3333–48.

Leiserowitz, A., E. Maibach, C. Roser-Renouf, and N. Smith. *Climate Change in the American Mind: Americans' Global Warming Beliefs and Attitudes in April, 2013*. Yale University and George Mason University. New Haven: Yale Project on Climate Change Communication. http://environment.yale.edu/climate/files /ClimateBeliefsJune2010.pdf.

Livezey, Robert E., K. Vinnikov, M. Timofaveyva, R. Tinker, and H. van den Dool. "Estimation and Extrapolation of Climate Normals and Climatic Trends." *Journal of Applied Meteorology and Climatology* 46 (November 2007): 1759–76.

Lueck, Michelle. "Environment Migration: Vulnerability, Resilience, and Policy Options for Internally Displaced Persons in the United States." In *SOURCE 15/2011*, edited by M. Leighton and X. Shen. Bonn, Germany: United Nations University Institute for Environment and Human Security, 2011.

McIntosh, Molly. "Measuring the Labor Market Impacts of Hurricane Katrina." Paper presented at Allied Social Science Associations Annual Convention, December 2007.

Mann, Charles C. *1493: Uncovering the New World Columbus Created*. New York: Vintage, 2012.

Marine Conservation Institute and Mission Blue. *Seastates: How Well Does Your State Protect Your Coastal Waters?* 2013.

Maryland Department of the Environment. "Executive Summary." Maryland's Greenhouse Gas Reduction Plan. July 2013.

http://climatechange.maryland.gov/publications/maryland-s-greenhouse-gas
-reduction-plan-executive-summary/.

McFarlane, Benjamin J., and staff, Hampton Roads Planning District Commission. "Phase III, Sea Level Rise in Hampton Roads, Virginia." *Climate Change in Hampton Roads.* July 2012.

McLachlan, Jason S., James S. Clark, and Paul S. Manos. "Molecular Indicators of Tree Migration Capacity under Rapid Climate Change." *Ecology* 86, no. 8 (2005): 2088–98.

McNulty, Steve. "George Washington National Forest, Case Study." Department of Agriculture, U.S. Forest Service. www.sgcp.ncsu.edu:8090/PlanningReport. aspx.

Meehl, Gerald A., Julie M. Arblaster, and Grant Branstator. "Mechanisms Contributing to the Warming Hole and the Consequent U.S. East-West Differential of Heat Extremes." *Journal of Climate* 25 (2012): 6394–408.

Meehl, Gerald A., Julie M. Arblaster, John T. Fasullo, Aixue Hu, and Kevin E. Trenberth. "Model-Based Evidence of Deep-Ocean Heat Uptake during Surface-Temperature Hiatus Periods." *Nature Climate Change* 1 (September 18, 2011).

Mitchell, M., C. Hershner, J. Herman, D. Schatt, E. Eggington, and S. Stiles. *Recurrent Flooding Study for Tidewater Virginia.* Virginia Senate Document No. 3. Richmond. Report. 2013.

Nash, Stephen. "An Acidifying Estuary? The 'Other CO_2 Problem.'" *Chesapeake Quarterly* 11, no. 1 (March 2012): 2–7.

———. *Blue Ridge 2020: An Owner's Manual.* Chapel Hill: University of North Carolina Press, 1999.

———. "Double Vision — Climate Change Comes to the Mountains." *Blue Ridge Country,* November–December 2008, 52–61.

———. "'Massive Trees' of Coral Thrive Where the Sun Doesn't Shine." *Washington Post,* June 11, 2013, E1.

———. "Supercomputer Should Improve Climate and Weather Forecasts." *Washington Post,* May 29, 2012, E1.

———. "Waterworld." *New Republic,* September 24, 2010, 6.

———. "Wetlands, Icecaps, Unease: Sea-Level Rise and Mid-Atlantic Shorelines." *BioScience* 58, no. 10 (November 2008): 919–23.

National Capital Planning Commission. "Urban Development Has Increased." *Report on Flooding and Stormwater in Washington, D.C.* January 2008.

National Center for Atmospheric Research. "NCAR Selects IBM for Key Components of New Supercomputing Center." Press release. November 7, 2011.

National Highway Traffic Safety Administration. "Summary of Fuel Economy Performance," April 28, 2011.

National Oceanic and Atmospheric Administration (NOAA). Earth System Research Laboratory, Global Monitoring Division. "Atmospheric CO_2 at Loa Observatory." www.esrl.noaa.gov/gmd/ccgg/trends/.

———. "Flood Event of 10/15/1942–10/17/1942." www.erh.noaa.gov/marfc/Rivers /FloodClimo/MARFCHistoricFloodEvents/1900sFloods/1942/1942-October15 .pdf. http://chesapeakebay.noaa.gov/oysters/oyster-reefs.

———. National Climate Assessment and Development Advisory Committee. *Draft National Climate Assessment.* January 2013.

———. National Climate Data Center. *Climate.Gov, Science & Information for a Climate-Smart Nation.* "Hottest.Month.Ever… Recorded." www.climate.gov /news-features/featured-images/hottestmonthever…-recorded.

———. National Climate Data Center. "Global Summary Information — July 2013–July 2013 global temperatures sixth highest on record." www.ncdc.noaa .gov/sotc/.

———. National Climate Data Center. "Virginia?Temperature? 60-Month Period Ending in December, 1899–2012, 60 month average, all divisions." www.ncdc .noaa.gov/cag/time-series/us.

——— "National Overview–February 2012." www.ncdc.noaa.gov/sotc/national .2012/2.

———. Northeast Fisheries Science Center. *Science Spotlight,* September 18, 2012. www.nefsc.noaa.gov/press_release/2012/SciSpot/SS1209.

———. "Oyster Reefs." http://chesapeakebay.noaa.gov/oysters/oyster-reefs.

———. "Tides and Currents." http://tidesandcurrents.noaa.gov/tides05/tab2ec2c .html.

———. "Virginia Tropical Cyclone Climatology." www.hpc.ncep.noaa.gov/ research/roth/vaclimohur.htm.

National Research Council. *Abrupt Impacts of Climate Change: Anticipating Surprises.* Washington, D.C.: National Academies Press, 2013.

———. Committee on the Development of an Integrated Science Strategy for Ocean Acidification. Monitoring, Research, and Impacts Assessment. *A National Strategy to Meet the Challenges of a Changing Ocean.* Washington, D.C.: National Academies Press, 2008. Summary at "Key Messages." http:// dels.nas.edu/Report/Ocean-Acidification-National-Strategy/12904.

———. Committee on Stabilization Targets for Atmospheric Greenhouse Gas Concentrations. *Climate Stabilization Targets: Emissions, Concentrations, and Impacts over Decades to Millennia.* Washington, D.C.: National Academies Press, 2011.

National Science and Technology Council. *Scientific Assessment of the Effects of Global Change on the United States — A Report of the Committee on Environment and Natural Resources.* May 2008.

The Nature Conservancy. "Virginia — Places We Protect." www.nature.org /ourinitiatives/regions/northamerica/unitedstates/virginia/placesweprotect /index.htm.

Nicholls, R. J., S. Hanson, C. Herweijer, N. Patmore, S. Hallegatte, Jan Corfee-Morlot, Jean Chateau, R. Muir-Wood, et al. "Organization for Economic Cooperation and Development Ranking of the World's Cities Most Exposed to Coastal Flooding Today and in the Future — Executive Summary." 2007. oecd.org/env/cc/39721444.pdf.

Ni, Xijun, Daniel L. Gebo, Marian Dagosto, Jin Meng, Paul Tafforeau, John J. Flynn and K. Christopher Beard. "The Oldest Known Primate Skeleton and Early Haplorhine Evolution." *Nature* 498 (June 6, 2013): 60–64.

Nye, Janet, Jason Link, Jonathan A. Hare, and William J. Overholtz. "Changing Spatial Distribution of Fish Stocks in Relation to Climate and Population Size on the Northeast United States Continental Shelf." *Marine Ecology Progress Series* 393 (2009): 111–29.

Old Dominion University Economic Forecasting Project. *The State of the Region — Hampton Roads.* 2009.

Oreskes, Naomi. "The Scientific Consensus on Climate Change." *Science* 306, no. 5702 (December 3, 2004): 1686.

Oreskes, Naomi, and Erik M. Conway. *Merchants of Doubt — How a Handful of Scientists Obscured the Truth on Issues from Tobacco Smoke to Global Warming.* New York: Bloomsbury Press, 2010.

Paul, Aaron. *The Economic Benefits of Natural Goods and Services — A Report for the Piedmont Environmental Council.* Yale School of Forestry and Environmental Studies. November 2011.

Pielke, Roger A., Jr. *The Honest Broker: Making Sense of Science in Policy and Politics.* Cambridge: Cambridge University Press, 2007.

Pilkey, Orrin, and Rob Young. *The Rising Sea.* Covelo, Calif.: Island Press, 2011.

Port of Richmond. "City of Richmond Marine Facility Lessee: Virginia Port Authority — Facts." www.richmondgov.com/PortOfRichmond/index.aspx.

Pyke, C. R., R. G. Najjar, M. B. Adams, D. Breitburg, M. Kemp, C. Hershner, R. Howarth, et al. *Climate Change and the Chesapeake Bay: State-of-the-Science Review and Recommendations. A Report from the Chesapeake Bay Program Science and Technical Advisory Committee.* STAC Publication #08-004. 2008.

Randers, Jorgen, and Paul Gilding. "The One Degree War Plan." *Journal of Global Responsibility* 1, no. 1 (2010): 170–88.

Reisen, W. K., R. M. Takahashi, B. D. Carroll, and R. Quiring. "Delinquent Mortgages, Neglected Swimming Pools, and West Nile Virus, California." *Emerging Infectious Disease Journal* (November 2008). wwwnc.cdc.gov/eid /article/14/11/08-0719.htm.

Revkin, Andrew. "On Issues Like Climate Change, Can Urgency and Patience Coexist?" *Dot Earth* (blog). *New York Times,* December 24, 2012. http://dotearth.blogs.nytimes.com/2012/12/24/urgency-and-patience-required -when-dealing-with-wicked-issues-like-climate-change/?_php=true&_type =blogs&_r=0.

Reynolds, R. W., and T. M. Smith. "A High-Resolution Global Sea Surface Temperature Climatology." *Journal of Climate* 8 (1995): 1571–83.

Richmond, G. M., and D. S. Fullerton. "Summation of Quaternary Glaciations in the United States of America." *Quaternary Science Reviews* 5 (1986): 183–96.

Riehl, Herbert, and Dave Fultz. "Jet Stream and Long Waves in a Steady Rotating-Dishpan Experiment: Structure of the Circulation." *Quarterly Journal of the Royal Meteorological Society* 83, no. 356 (April 1957): 215–31.

Robbins, Jim. "What's Killing the Great Forests of the American West?" *Environment360,* March 15, 2010. http://e360.yale.edu/feature/whats_killing _the_great_forests_of_the_american_west/2252/.

Roberts, J. Murray, Andrew Wheeler, Adre Freiwald, and Stephen Cairns. *Cold-Water Corals: The Biology and Geology of Deep-Sea Coral Habitats.* Cambridge: Cambridge University Press, 2009.

Rogers, Karilon L. "People, Planet, Profit — Combining Positive Energy and Social Responsibility: Tom Dingledine (MBA '78) Leads the Business of Change." *Wake Forest Magazine,* September 2007, 36–37.

Rothschild, B. J., J. S. Ault, P. Goulletquer, and M. Heral. "Decline of the Chesapeake Bay Oyster Population: A Century of Habitat Destruction and Overfishing." *Marine Ecology Progress Series* 111 (August 1994): 30–39.

Sallenger, Asbury H., Jr., Kara S. Doran, and Peter A. Howd. "Hotspot of Accelerated Sea-Level Rise on the Atlantic Coast of North America." *Nature Climate Change,* June 24, 2012.

Sandell, Clayton. ABC News. "Senators to Exxon: Stop the Denial." http:// abcnews.go.com/Technology/story?id=2612021&page=1.

Schmidt, Gavin. NASA Goddard Institute for Space Studies and Center for Climate Systems Research. "How Earth System Models Are Reshaping the Science/ Policy Interface." Alliance program, Columbia University. October 2011.

Shabecoff, Philip. "E.P.A. Proposes Rules to Curb Warming." *New York Times,* March 14, 1989.

Shepherd, Andrew, and Erik R. Ivins. "A Reconciled Estimate of Ice-Sheet Mass Balance." *Science* 338, no. 6111 (November 30, 2012): 1183–89.

State Climate Office of North Carolina. "Staff." www.nc-climate.ncsu.edu/office /staff.html.

Stein, Walter J. *California and the Dust Bowl Migration.* Westport, Conn.: Greenwood Press, 1973.

Stephenson, N. L., A. J. Das, R. Condit, S. E. Russo, P. J. Baker, N. G. Beckman, D. A. Coomes, et al. "Rate of Tree Carbon Accumulation Increases Continuously with Tree Size." *Nature,* January 15, 2014. Online.

Stoner, Anne M. K., Katharine Hayhoe, Xiaohui Yang, and Donald J. Wuebbles. "An Asynchronous Regional Regression Model for Statistical Downscaling of Daily Climate Variables." *International Journal of Climatology* (September 2013): 2473–94.

Street, Michael W., Anne S. Deaton, William S. Chappell, and Peter D. Mooreside. *North Carolina Coastal Habitat Protection Plan.* North Carolina Department of Environment and Natural Resources, 2005.

Titley, David. Blue Planet Forum. Lecture. December 2, 2010. Sponsors: Old Dominion University Office of Community Engagement, Chesapeake Bay Foundation, National Oceanic and Atmospheric Administration, Nauticus sea exhibit facility.

Titus, J. G., K. Eric Anderson, Donald R. Cahoon, Dean B. Gesch, Stephen K. Gill, Benjamin T. Gutierrez, E. Robert Thieler, and S. Jeffress Williams. *Coastal Sensitivity to Sea-Level Rise: A Focus on the Mid-Atlantic Region.* Washington D.C.: U.S. Climate Change Science Program and the Subcommittee on Global Change Research, U.S. Environmental Protection Agency, 2009.

Titus, J. G., and D. E. Hudgens, eds. *The Likelihood of Shore Protection along the Atlantic Coast of the United States.* Vol. 1, *Mid-Atlantic.* Washington, D.C.: U.S. Environmental Protection Agency, 2010.

Titus, J. G., D. E. Hudgens, C. Hershner, J. M. Kassakian, P. R. Penumalli, M. Berman, and W. H. Nuckols. "Virginia." In *The Likelihood of Shore Protection along the Atlantic Coast of the United States,* edited by Titus and Hudgens, vol. 1, *Mid-Atlantic.* Washington, D.C.: U.S. Environmental Protection Agency, 2010.

Titus, J. G., D. E. Hudgens, D. L. Trescott, M. Craghan, W. H. Nuckols, C. H. Hershner, J. M. Kassakian, et al. "State and Local Governments Plan for Development of Most Land Vulnerable to Rising Sea Level along the US Atlantic Coast." *Environmental Research Letters* 4, no. 4 (October–December 2009). http://iopscience.iop.org/1748-9326/4/4/044008/fulltext/.

Titus, J. G., and U.S. Environmental Protection Agency. *Climate Ready Estuaries Program, Rolling Easements.* June 2011.

Tollefson, Jeff. "The Case of the Missing Heat." *Nature* 505 (January 16, 2014): 276–78.

United Nations Department of Economic and Social Affairs. "Greenhouse Gas Emissions per Capita." July 2010. http://unstats.un.org/unsd/environment/air_greenhouse_emissions.htm.

United Nations Sustainable Development Knowledge Platform. "Oceans — Facts and Figures." Undated. www.un.org/en/sustainablefuture/oceans.shtml.

U.S. Department of Agriculture. "National Agricultural Statistics Service Crop Production 2012 Summary." January 2013.

U.S. Department of Commerce. Census Bureau. "State & County Quick Facts, Richmond City, Virginia." http://quickfacts.census.gov/qfd/states/51/51760.html.

———. National Oceanic and Atmospheric Administration (NOAA). See National Oceanic and Atmospheric Administration (NOAA).

U.S. Department of Energy. Energy Information Administration. "How Much Electricity Does an American Home Use?" www.eia.gov/tools/faqs/faq.cfm?id=97&t=3.

———. Energy Information Administration. "Most States Have Renewable Portfolio Standards." Today in Energy, February 3, 2012. www.eia.gov/todayinenergy/detail.cfm?id=4850.

———. Energy Information Administration. "State-Level Energy-Related Carbon Dioxide Emissions, 2000–2010." May 2013.

———. Energy Information Administration. "Virginia Profile." www.eia.gov/state/analysis.cfm?sid=VA&CFID=12344435&CFTOKEN=5872a78a21abd773-F33B8ED9-237D-DA68-243F3AFA7013FADE&jsessionid=8430314b611a86224b0f3b561f114c566383.

———. Energy Information Administration, Office of Energy Analysis. "International Energy Outlook 2013 with Projections to 2040." July 2013.

U.S. Department of Labor. Bureau of Labor Statistics. "CPI Inflation Calculator." www.bls.gov/data/inflation_calculator.htm.

U.S. District Court for the District of Vermont. "Motion to intervene of Dr. Patrick J. Michaels." July 6, 2007. In Green Mountain Chrysler Plymouth Dodge Jeep, et al. vs. George Crombie; Greenpeace, Inc., Intervenor, 1–4.

U.S. Environmental Protection Agency. *Integrated Science Assessment for Particulate Matter (Final Report).* 2009.

———. "Overview of EPA's Revisions to Air Quality Standards for Particle Pollution (Particulate Matter)." epa.gov/pm/2012/decfsoverview.pdf/.

———. "Total CO_2 from Fossil Fuels by Sector for States/Regions" (2009). www.epa.gov/reg3artd/globclimate/ccghg.html#Total%20Emissions.

U.S. Forest Service. Southern Region. *Draft Revised Land and Resource Management Plan.* George Washington National Forest. April 2011.

U.S. Government Accountability Office. *Report to the Chairman, Committee on Commerce, Science, and Transportation, U.S. Senate: Electricity — Significant Changes Are Expected in Coal-Fueled Generation, but Coal Is Likely to Remain a Key Fuel Source.* October 2012.

Virginia Department of Conservation and Recreation. "The Natural Communities of Virginia Classification of Ecological Community Groups." www.dcr.virginia.gov/natural_heritage/nchome.shtml.

Virginia Department of Environmental Quality. "Gross Emissions of Greenhouse Gases in Virginia." Unpaginated.

———. VIRGINIA Draft 305(b)/303(d). Water Quality Integrated Report to Congress and the EPA Administrator for the Period January 1, 2005, to December 31, 2010. Richmond, Virginia. March 2012.

Virginia Department of Forestry. "Forest Facts." www.dof.virginia.gov/resinfo /forest-facts.shtml.

Virginia Department of Game and Inland Fisheries. *Virginia's Comprehensive Wildlife Conservation Strategy.* Virginia Department of Game and Inland Fisheries. Richmond. 2005.

Virginia Department of Health. Office of Epidemiology, Department of Surveillance and Investigation. "Table 2. Ten-Year Trend in Number of Reported Cases of Notifiable Diseases in Virginia, 2002–2011." www.vdh.virginia.gov /Epidemiology/Surveillance/SurveillanceData/ReportableDisease/.

Virginia Department of Mines, Minerals and Energy. "Coal in Virginia Fact Sheet." www.dmme.virginia.gov/dgmr/coal.shtml.

Virginia Institute of Marine Science. *Land-Coast Connections and Climate Change: Carbon Cycling in Chesapeake Bay and Its Watershed.* October 2008. www.vims.edu/research/units/programs/icccr/_docs/land_coastal _connections.pdf.

———. Center for Coastal Resources Management. *Regulatory Fidelity to Guidance in Virginia's Tidal Wetlands Program — Final Report.* December 2012.

Virginia Marine Products Board. "About Virginia Seafood." www.virginia seafood.org/dive-in/about/.

Walsh, Margaret, Peter Backlund, Anthony Janetos, and David Schimel. *The Effects of Climate Change on Agriculture, Land Resources, Water Resources, and Biodiversity in the United States. Report by the U.S. Climate Change Science Program and the Subcommittee on Global Change Research.* 2008.

Walthall, C. L., et al. *Climate Change and Agriculture in the United States: Effects and Adaptation.* Technical Bulletin 1935. Washington, D.C.: U.S. Department of Agriculture, 2013.

Watanabe, Masahiro, Youichi Kamae, Masakazu Yoshimori, Akira Oka, Makiko Sato, Masayoshi Ishii, Takashi Mochizuki, and Masahide Kimoto. "Strengthening of Ocean Heat Uptake Efficiency Associated with the Recent Climate Hiatus." *Geophysical Research Letters* 40 (June 28, 2013): 3175–79.

Watts, Anthony. "About." Watts Up with That? website. http://wattsupwiththat .com/about-wuwt/about2/.

Watts, W. A. "Late Quaternary Vegetation of Central Appalachia and the New Jersey Coastal Plain." *Ecological Monographs* 49, no. 4 (December 1979): 427–69.

Webster, Don, and Don Meritt. "The Future of Oysters in Chesapeake Bay—Different Paths to Restoration." *Maryland Aquafarmer online,* Summer 2001. www.mdsg.umd.edu/sites/default/files/files/Maryland%20Aquafarmer%20-%20Summer%202001.pdf.

Wharton, J. *The Bounty of the Chesapeake: Fishing in Colonial Virginia.* Virginia 350th Anniversary Celebration. Williamsburg, Va., 1957. Unpaginated.

Wieseltier, Leon. "Who by Water." *New Republic,* December 6, 2012, 52.

Willis, C. G., B. R. Ruhfel, R. B. Primack, A. J. Miller-Rushing, J. B. Losos, et al. "Favorable Climate Change Response Explains Non-Native Species' Success in Thoreau's Woods." *PLOS ONE* (2010): e8878.

Yang, Ailun, and Yiyun Cui. *Global Coal Risk Assessment: Data Analysis and Market Research.* World Resources Institute working paper. November 2012.

Zhu, Kai, Christopher W. Woodall, and James S. Clark. "Failure to Migrate: Lack of Tree Range Expansion in Response to Climate Change." *Global Change Biology* 18 (2012): 1042–52.

ILLUSTRATION SOURCES

Fig. 1: National Climate Data Center/National Oceanic and Atmospheric Administration map at www.ncdc.noaa.gov/monitoring-references/maps/us -climate-divisions.php.

Figs. 2-8: Robert Livezey, based on data from the National Climate Data Center, National Oceanic and Atmospheric Administration.

Fig. 9: Dana Nuccitelli, Skeptical Science website, adapted from Intergovernmental Panel on Climate Change, *Emissions Scenarios,* 8–10; International Energy Agency, "CO$_2$ Emissions from Fuel Combustion Highlights."

Fig. 10: Chris Zganjar, climate change applications developer, The Nature Conservancy. Zganjar used a median (50th percentile) of the combination of sixteen General Circulation Models to create this figure, using model runs based on global climate projections from the World Climate Research Programme's Coupled Model Intercomparison Project phase 3 (CMIP3) multimodel dataset, which was referenced in the Intergovernmental Panel on Climate Change's *Assessment Report 4.* The emissions scenarios used are A2 and B1.

Figs. 11–13: Sharmistha Swain and Katharine Hayhoe, based on Stoner et. al, "An Asynchronous Regional Regression."

Fig. 14: Chris Zganjar, climate change applications developer, The Nature Conservancy. Of fifteen general circulation models of the A2 emissions analysis, the maps show the values of the highest and the lowest model.

Fig. 15: Melissa Clark, The Nature Conservancy.

Fig. 16: Army Corps of Engineers, http://corpsclimate.us/ccaceslcurves.cfm.

Figs. 17–21. Mitchell et al., *Recurrent Flooding Study,* 74 (fig. 17), 75 (fig. 18), 76 (fig. 19), 77 (fig. 20), 73 (fig. 21).

Fig. 22: Larry P. Atkinson, Professor of Oceanography, Center for Coastal Physical Oceanography, Old Dominion University.

Fig. 23: Virginia Department of Environmental Quality, "Gross Emissions of Greenhouse Gases in Virginia."

Fig. 24: Adam Voiland, NASA Earth Observatory, from Leibensperger et al., "Climatic Effects of 1950–2050 Changes in US Anthropogenic Aerosols."

Fig. 25: Katharine Hayhoe, director, Climate Science Center, Texas Tech University, using MAGICC, an energy balance model in which climate sensitivity is prescribed, based on the IPCC range of sensitivities.

INDEX